Levi N. Beebe

First Steps among Figures

A Drill Book in the fundamental Rules of Arithmetic

Levi N. Beebe

First Steps among Figures
A Drill Book in the fundamental Rules of Arithmetic

ISBN/EAN: 9783337158002

Printed in Europe, USA, Canada, Australia, Japan

Cover: Foto ©Paul-Georg Meister /pixelio.de

More available books at **www.hansebooks.com**

FIRST STEPS
AMONG FIGURES.

A Drill Book in the Fundamental Rules of Arithmetic.

TEACHERS' EDITION.

BY

LEVI N BEEBE,

CANANDAIGUA, N. Y.

SIXTH EDITION, ENLARGED AND CAREFULLY REVISED.

SYRACUSE, N. Y.:
C. W. BARDEEN, PUBLISHER.
1881.

PREFACE TO TEACHERS' EDITION.

In putting this work before the public the author disclaims any ambitious schemes or "great expectations," but he wishes to have the book for the use of his assistant teachers both as to methods and examples. The author has used some parts of it for many years and feels confident that excellent results may be obtained by using it.

The aim of the book is to give so much practice as to fix each *method* in the pupil's mind, rather than to deal with the philosophy of each operation, leaving any teacher who believes that no step should be taken unless the pupil understands the reasoning process by which that step may be reached, to give it in his own way. It is possible that a few who see this book may have found that 7 times 8 are 56 by actual addition, yet those who have never added it may know the fact just as well for all practical purposes.

If no one were to eat until he understood how food nourishes the system there would be a deal of hunger in the world.

This book deals only with the fundamental rules of arithmetic. The intention is that they shall be so thoroughly mastered that much less time will be required for the remainder of the subject of arithmetic than would otherwise be needed.

The teacher is to use the Teachers' Edition for one to two years before the pupil has advanced enough to use the Pupils' Edition or in fact any book on arithmetic. It is recommended that teachers begin to teach numbers as given in the first part of this book after pupils who have the alphabet and words to learn have been in school four to six months.

In each new operation the examples are very easy ; as more problems are given they gradually increase in difficulty. By teaching the four operations of addition, subtraction, multiplication, and division, from the first, the examples are of such a kind as to compel some thoughtfulness on the part of the pupil.

Much pains has been taken to make examples of a sort to interest the youngest pupils.

Those teachers who wish to teach only addition and subtraction at first can designate those examples involving multiplication or division by some mark, and omitting them may return to them afterwards and so secure the variety of examples so essential to a pupil's real progress. It has been found, however, by actual trial that pupils may learn the four operations from the first without serious difficulty.

The first pages are devoted to what is known as the "Grube Method." If the teacher prefers it, the schedules may be omitted, and, in passing through the first time, the multiplication and division also, as before stated.

The author hopes that teachers into whose hands this work may come will give it a thorough examination. Special attention is called to the treatment of numeration and notation. The examples are not all given in one place, to be forgotten, but are so placed as to review the subject often.

Attention is called to the examples for rapid solving and the illustration of the easy examples given under each rule. Also to the method of teaching long division and to the definition of addition.

The method of teaching the addition, subtraction, multiplication and division tables is believed to be entirely new, so far as being published is concerned. The author discovered and used the method about ten years ago, and in his school has found it invaluable.

To hear a recitation of a large class in tables and make the questions to each pupil promiscuous, and yet full enough to satisfy the teacher that the pupil has a thorough knowledge of the tables gone over, is not only very wearying to the teacher but is exceedingly difficult also. By the old method a pupil frequently acquires the habit of saying the table from the beginning to find the result of any combination, as 7 times 6.

To enable any one to make new series like those here given, I insert the method. The following is for 9's and review. In the given lines of figures there is one more figure in the upper line than in the lower one:

$$\begin{array}{ccccccc} 3 & 4 & 5 & 6 & 7 & 8 & 9 \\ 4 & 5 & 6 & 7 & 8 & 9 & \end{array}$$

If the upper line be written several times and the lower line in the same way as follows:

$$\begin{array}{cccccccccccccc} 3 & 4 & 5 & 6 & 7 & 8 & 9 & 3 & 4 & 5 & 6 & 7 & 8 & 9 \\ 4 & 5 & 6 & 7 & 8 & 9 & 4 & 5 & 6 & 7 & 8 & 9 & 4 & 5 \\ 3 & 4 & 5 & 6 & 7 & 8 & 9 & 3 & 4 & 5 & 6 & 7 & 8 & 9 \\ 6 & 7 & 8 & 9 & 4 & 5 & 6 & 7 & 8 & 9 & 4 & 5 & 6 & 7 \\ 3 & 4 & 5 & 6 & 7 & 8 & 9 & 3 & 4 & 5 & 6 & 7 & 8 & 9 \\ 8 & 9 & 4 & 5 & 6 & 7 & 8 & 9 & 4 & 5 & 6 & 7 & 8 & 9 \end{array}$$

The first 9 in the lower line comes one place before the 9 in the upper line; the second 9 in the lower line comes two places before the second 9 in the upper line, and so on until it has been under every figure in the upper line. If written farther, the series will be repeated as shown above, where 4 and 3 occur again at the end as they did at the beginning. The upper line of figures must be written one less number of times than there are figures in it.

This series may be used for addition or multiplication, thus: 4 and 3 are 7, 5 and 4 are 9, &c., or 4 times 3 are 12, 5 times 4 are 20, &c.

PREFACE TO TEACHERS' EDITION. 7

This arrangement is objectionable for most of the results in addition vary only by 2 or 4 and are not as promiscuous as they should be. By disarranging the upper line of figures we have 6, 9, 5, 8, 4, 7, 3. Re-writing this for the upper line and writing the lower line as before we have

```
6 9 5 8 4 7 3 6 9 5 8 4 7 3
4 5 6 7 8 9 4 5 6 7 8 9 4 5
6 9 5 8 4 7 3 6 9 5 8 4 7 3
6 7 8 9 4 5 6 7 8 9 4 5 6 7
6 9 5 8 4 7 3 6 9 5 8 4 7 3
8 9 4 5 6 7 8 9 4 5 6 7 8 9
```

which, like the other series, contains every combination between 4 and 3 and 9 and 9 inclusive, and none repeated except by inversion as $4+5$ and $5+4$; but unlike that series it is entirely promiscuous.

To make a series for subtraction, write the series as above, and write the sums of the numbers above, thus:

```
10 14 11 15 12 16  7 11 15 12 16 13. &c.
 6  9  5  8  4  7  3  6  9  5  8  4  7 3 &c.
 4  5  6  7  8  9  4  5  6  7  8  9   &c.
```

Then copy for the minuends the upper numbers, and for the subtrahends the lower ones and the series becomes

```
10 14 11 15 12 16  7 11 15 12 16 13  &c.
 4  5  6  7  8  9  4  5  6  7  8  9  &c.
```

In this book those for subtraction have been still further disarranged so that the results will not be 6 9 5 8 4 7 3 and so on lest the pupils notice it and recite that instead of subtracting. For division find the products for the upper line instead of the sums.

For division with remainders, *which is an excellent preparation for short division*, after having written the products above as before, add to each one of them a number less than the lower number in that column and write for the upper line these sums and for the lower line the lowest line of figures.

In the first series of division with remainders, the remainders are very small, that it may be as easy as possible.

In the first series the combinations do not go as far as 9— that is 2 and 9, 9 times 2, &c., but only to combinations of 2 3 4 5 and 6 with 2 3 4 5 and 6. The examples which follow immediately after the learning of any table involve only what is contained in the table.

It is believed there is a very large amount of work for practice, both in the Pupils' Edition and in the Teachers' Edition, more than twice as much as in other works of the kind. The greatest care has been taken that they may proceed from the easiest to those involving every difficulty which pupils should meet at the age for which this book is designed.

As anything is learned it is immediately put into use.

LEVI N. BEEBE.

Canandaigua, N. Y., July, 1877.

PART I.

ONE.

(See Appendix, page 185.)

1. Be sure the pupil has the idea of *one* thing in distinction from *two* or more things.

Illustrate by objects as much as possible, using small sticks, or square blocks ¾ inch square and ¼ inch thick of different colors, or bright cents. Only ten of each are needed and if the teacher has *all* these he can add interest to the exercises.

An abacus, or numeral frame, is almost indispensable as a further help.

Show the pupil that taking one article (as a bean, a cent or a block) one time, or putting it into a box or upon a book or table makes one article there, which is the interpretation of "once one is one." Let the *pupil* place the article, and thus impress his mind more thoroughly with the idea once 1 is 1, written $1 \times 1 = 1$.*

---

*In this book the multiplier is uniformly placed on the right of the sign $\times$: thus 2 times one are 2 will be written $1 \times 2 = 2$.

See Appendix, pp. 185-192.

2. The idea of division may be taught in the following way: The teacher may place a pile of 2 blocks on a table or book and ask the pupil "How many times one block have I in this pile?" Pupil: "Two times." Teacher: "One block in two blocks how many times?" Pupil: "Two times." Teacher: "One in two how many times?" Pupil: "Two times." (The pupil may use in each answer the word "twice" instead of the words "two times.") Teach that this is written $2 \div 1 = 2$ and should be read by the youngest pupils, 1 in 2 twice.

Schedule:

$1 \times 1 = 1$. (Read once one is one.)

$1 \div 1 = 1$. (Read one in one, once.)

What can you find once in the school room, in your pocket, on your face, at home? &c.

What is there that moves on one wheel?

TWO.

3. Teach in counting that the second of two things is not of itself *two*, but *one*.

In teaching number and in operations on numbers use objects for some time — at least

---

*The schedules being written on the blackboard, the pupils are to be taught to read them, and eventually to make them themselves.

three months to six months, until the pupil is thoroughly familiar with the composition of numbers.

4. Teach pupils to count to 11 and continue to teach counting daily until the pupil can count 100.

Schedule:

$1 + 1 = 2$. (Read one and one are two.)
$1 \times 2 = 2$. (Read twice one are two.)
$2 - 1 = 1$. (Read one from two leaves one.)
$2 \div 1 = 2$. (Read one in two twice.)

2 is one more than what number?
1 is one less than what number?
2 is the double of what number?
2 is twice what number?
1 is one-half of what number?
1 and 1 are*?   1 from 2 leaves?

(1 from 2 leaves 1, because 1 and 1 are 2.)

Mary has 2 sticks of candy; she gives away 2 sticks; how many sticks has she left? 2 from 2 leaves?

Henry had 2 marbles; he gives none away: how many has he? Nothing from 2 leaves?

---

*The teacher will supply the words "what number" or "how much" in such examples, according to the sense.

Show the pupils that 1 block placed on the table, and then another, make 2 blocks there, hence 2 is 2 times 1.

What is there that moves on two wheels?

Hold up two fingers.

What have you on your head of which there are 2 and only 2? In the school room? At home? &c.

What animals walk on 2 legs?

5. What is ½ of an apple? (Let the pupil take an apple and cut it into halves and ask him what one piece is called. Show him that if he takes one-half of the apple there is left as much as he takes.) What is ½ of 2 apples? Placing 2 apples on the table, let one of the pupils take half of them by leaving as much as he takes.

(Vary the exercise by taking ½ of a stick of candy, ½ of 2 sticks, &c.)

6. 2 is the double of what number? Of what number is 1 one-half?

What number must I double to get 2?

I know a number that is 1 more than 1; what number is it?

7. What number must be added to 1 to get 2? Fred had 2 dimes and bought peaches with 1 dime. How many dimes had he left?

(No analysis of these examples is expected; simply a prompt answer.)

A slate pencil costs 1 cent, how much will 2 slate pencils cost?

Charles* had a marble, and his sister had twice as many. How many did she have?

How many slate pencils can you buy for 2 cents?

How many 2-cent stamps can you buy for 2 cents? How many 1-cent stamps?

(Both these and the following examples should be gone over many times, taking them in a different order each time and often giving them promiscuously.)

8. Teach that there are 2 pints in a quart by pouring a pint cup full of water twice into a quart cup.

What cost a quart of milk at 1 cent a pint?

THREE.

Schedule:

9. Measuring by 1.

$$\begin{array}{l} 1\ 1\ 1\ \ 3. \\ 1\ 1\ \ 1+1+1=3. \\ 1\ 1\ \ 1\times 3=3. \qquad\qquad [=1. \\ 1\ 1\ \ 3-1-1=1,\ \text{for}\ 3-1=2\ \text{and}\ 2-1 \\ \overline{3}\ \ 3\div 1=3. \end{array}$$

---

*In such examples it will interest the class to use *their* names instead of those given.

Measuring by 2.

  1 1 2   2+1=3.
     1 1   2×1+1=3. (To be read once 2
  ――― ―   and 1 are 3, or once 2 plus 1
  111 3   are 3.)

          3−2=1, 3−1=2. (To be read 2 from 3 leaves 1 and 1 from 3 leaves 2.)

          3÷2=1 (and 1 rem.) (To be read 2 in 3 once and 1 remainder.)

The pupils should read these schedules many times each, until they are familiar with the language.

10. Illustrate by a pile of 3 blocks. How many times have I 2 blocks in the pile? Once. Take them away once then. How many are left, or how many remain? One. 2 blocks in 3 blocks how many times? Once and 1 remainder. 2 in 3 how many times? Once and 1 remainder.

11. To be written on the blackboard for pupils to bring written with the answers to recitation.

  3÷1=?   1×2=?   3−1+1=?
  1+1=?   1+2=?   2+2=?

$3-2=?$ $3-1=?$ $2-2=?$
$2+1=?$ $2+1+1=?$ $1\times 2+1=?$
$2\div 2=?$ $3\div 2=?$ $3\div 3=?$

12. 3 is 1 more than?
 1 is 1 less than?
 3 is 2 more than?
 2 is 1 less than?
 2 is 1 more than?
 1 is 2 less than?
 3 is 3 times?

13. To illustrate tell a pupil to put one block or one cent on the desk and then another. Show the pupils that a block has been put upon the desk twice and that there are two blocks there; hence 2 times 1 block are 2 blocks; also 2 times 1 orange are 2 oranges, and 2 times 1 pencil are 2 pencils, &c. 2 times any one thing are two of those things. 2 times 1 are 2.

Show the pupils that 1 block taken 3 times or placed on a table 3 times makes 3 blocks there, hence 3 is 3 times 1.

14. This form of illustration may be used for any multiplication.

How many pints in a quart?

Teach pupils to write numbers as high as 20.

It may be well to teach the writing of 12 before 10 or 11. Show the pupil by the abacus or otherwise 12 objects and show him that they are 1 ten and 2 ones. Show him that we cannot write 12 by any *one* of our figures; then teach him about ten's place and one's place.

Do not use the word units for several weeks yet.

15. Teach pupils to count by 2's from 2 to 6 and back to 2, thus: 2, 4, 6. 6, 4, 2.

Explain that 1 and 1 are equal numbers, that is equal to each other; 1 and 2 as well as 2 and 3 are unequal numbers.

16. Give the pupils much practice in examples like the following: $3-1-1+1$? To be read, how many are 3 less 1 less 1 plus 1? or 3, subtract 1, subtract 1, add 1; or 3 minus 1 minus 1 plus 1.*

---

*These are to be read by the teacher, thus: 3, add 2, subtract 1, divide by 2, multiply by 3.

The examples may be read through and those who can answer raise the hand; the teacher call upon one most unlikely to be correct for the answer; if incorrect call upon another until the correct answer be given.

It may be best at first and perhaps often to have the result of the first step given by one pupil, the next step by the next, &c E. g. teacher, 3, add 2.   1st pupil says " 5." Teacher,

2−1+2−1×1= ?   1+2−1−1×2+1= ?
2+1−2×3−1+1= ? 1+1×1−1+2−1= ?

Read with as much rapidity as the class can follow silently and give the answer at the end, the rapidity being increased as the pupils have more practice.

17. From what number can you take one and have one left?

Count by 2's from 2 to 10.

What number is twice 1?

18. I write a number once, and again, to get 2; what number did I write twice?

How many cents must you have to buy a 3-cent stamp?

Mary had to get a pound of tea for $1; her mother gave her $3; how much money ought she to bring back?

Henry learned 1 line in his primer, and his

"subtract 1." 2d pupil says "4." Teacher, "divide by 2." 3d pupil "2." Teacher, "multiply by 3." 4th pupil "6." Call on an inattentive pupil at any step in these examples for the answer. Usually, the teacher reads the whole example and the pupils give only the final answer.

The foregoing examples are not written so as to be correct for solving from the written or printed form for in that case 3 — 2 × 2 would mean, take 2 × 2 from 3, but it is to be read: 3 subtract 2, multiply by 2.

sister learned 1 line more than he did; how many did she learn?

If 1 slate pencil cost 1 cent what will 3 slate pencils cost?

Anna found 3 roses in the garden; how can she divide them between her father and mother?

Can she give them an equal number?

How many roses must she have had in order to give her father 1, and her mother 1 also?

Count by 2's from from 2 to 12.

The counting may be sometimes in concert, oftener 1st pupil say 2; 2d pupil, 4; 3d, 6 and so on, and perhaps oftenest one pupil give the whole series.

Teach pupils to count 1st, 2d, 3d, &c.

FOUR.

Schedule:

20. Measuring by 1.

$$
\begin{array}{cc}
 & 1\ 1\ 1\ 1\quad 4. \\
1\ 1 & 1+1+1+1=4.\quad\text{(Because } 1+1=2, \\
 & 2+1=3,\ 3+1=4.) \\
1\ 1 & 1\times 4=4. \\
1\ 1 & 4-1-1-1=1,\text{ or } 4-1-1-1-1=0. \\
\dfrac{1}{4}\ \dfrac{1}{4} & 4\div 1=4.
\end{array}
$$

Measuring by 2.

```
 I  I   2.   2+2=4.
             2×2=4.
 I  I   2.   4−2=2, or 4−2−2=0.
─────────
IIII    4.   4÷2=2.
```

Measuring by 3.

```
 I  I  I   3.   3+1=4, 1+3=4.
                3+1+1=5.
       I   1.   4−3=1, 4−1=3.
─────────
IIII       4.   4÷3=1 (and 1 remainder.)
```

21.

3−2=?	2×1+2=?	4÷3=?
4×1=?	3÷1=?	3÷2=?
3−1=?	2+1=?	3+1=?
2+2=?	2×2=?	4−3=?
3÷2=?	2×1+1=?	4−1=?
4−2=?	3×1+1=?	4−4=?
1×3=?	4−2−2=?	
4÷2=?	4÷4=?	

22. Name animals with 4 legs; with 2 legs.

Name wagons and vehicles with 1 wheel; 2 wheels; 3 wheels; 4 wheels. Compare them. (For instance a wagon with 4 wheels has how many more wheels than one with 2 wheels? &c.)

23. 4 is 1 more than?
    1 is 1 less than?

2 is 1 more than?
3 is 1 less than?
1 is 1 more than?
1 is 2 less than?
2 is 1 less than?
2 is 2 more than?
2 is 2 less than?
4 is 2 more than?
1 is 1 more than? (Nothing.)
4 is 4 times?

24. Solve rapidly the following:

$2 \times 2 - 3 + 2 \times 1 + 1 - 2 \times 2 = ?$

$4 - 1 - 1 + 1 + 1 - 3 =$ how many less than 4?

$3 - 2 + 3 - 1 - 1 \times 2 - 1 =$ how many times 1?

$1 + 2 - 1 \div 2 + 2 - 1 =$ how many more than 2?

Teach to count by 2's from 2 to 20 and back to 2.

25. $1 \times 2 - 1 \times 3 - 2 - 2 =$ how many less than 3?

$3 - 2 + 1 \times 2 - 1 - 2 + 1 = ?$

$4 - 2 - 1 \times 3 - 1 \times 2 - 1 = ?$

26. What number must I double to get 4?
Of what number is 4 the double?
Of what number is 2 one-half?
What number can be taken twice from 4?
What number is 2 more than 1?
What number must I add to 2 to get 4?

What number is 1-2 of four?

How many less than 3 is the half of 4?

27. Minnie had 4 pinks which she neglected sadly; one day 1 of them withered, the second day another, and the following day 1 more. How many fresh ones had she then?

How many $'s are $2 + $2?

How many apples are 3 apples and 2 apples?

28. Teach that there are 4 quarts in 1 gallon, taking a gallon measure and filling it by pouring a quart measure full of water into it 4 times.

Nellie bought a gallon of milk; how many quarts did she buy?

She paid 1 dime for each quart; how many dimes did she pay for the gallon?

If 2 qts. of milk cost 2 di., can you get a gal. for 3 di.?

How much *can* you get for the 3 di.?

If I drink a quart of milk in 2 days, what part of a qt. do I drink in 1 da.?

29. William having 4 apples, ate half of them and one more, how many had he left?

What number is 1 more than half of 4?

Ann had 3 apples; she gave an equal number to her mother, father and brother; how many did she give each?

Sarah cut 1 apple into 2 equal pieces; what would you call one of the pieces?

Teach to count by 2's from 2 to 40 and back to 2.

By marking off paper or pasteboard, or better a thin board, and cutting, according to the following directions, an excellent aid in teaching notation and numeration may be obtained. By ruling both ways, mark off into squares 10 squares in a row and 21 rows, as shown below. Cut off one row or strip of 10 squares and then cut up the strip into single squares. Afterwards cut off 10 strips of 10 squares, which will leave a large square containing 10 rows of small squares with 10 squares in each row.

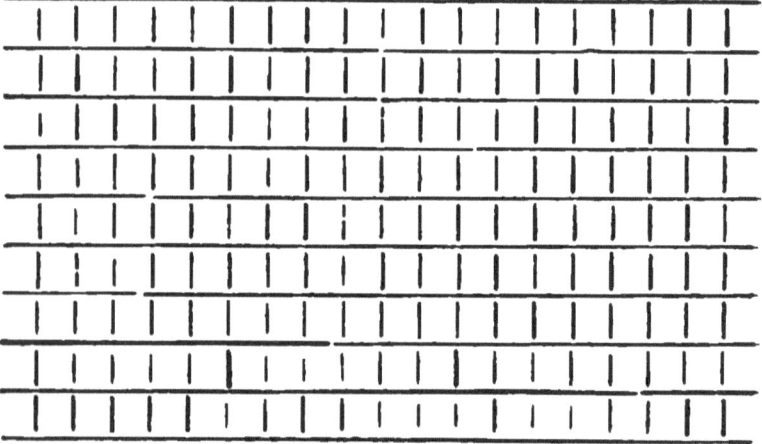

The 10 small squares cut up may be used to illustrate ones or units, and the strips, tens, while the large square will represent hundreds. With them it will be easy to show that 10 ones equal a ten, and ten tens one hundred, and the teacher will show where the ones, tens, and hundreds are written in numbers.

FIVE.
Schedule;

30. Measuring by 1.

$$\begin{array}{ll} & 1\ 1\ 1\ 1\ 1\quad 5. \\ \mathit{1}\ 1 & 1+1+1+1+1=5. \\ \mathit{1}\ 1 & 1\times 5=5. \\ \mathit{1}\ 1 & 5-1-1-1-1=1. \\ \mathit{1}\ 1 & 5\div 1=5. \\ \mathit{1}\ 1 & \\ \overline{5}\ \overline{5.} & \end{array}$$

Measuring by 2.

$$\begin{array}{lll} 1\ 1 & 2. & 2+2+1=5. \\ 1\ 1 & 2. & 2\times 2+1=5.\quad\text{(See note p. 23.)} \\ \underline{\ \ 1} & \underline{2.} & 5-2-2=1. \\ 11111 & 5. & 5\div 2=2\ (1\text{ remainder.}) \end{array}$$

By 3.

$$\begin{array}{lll} 1\ 1\ 1 & 3. & 3+2=5,\ 2+3=5. \\ \underline{1\ 1} & \underline{2.} & 3\times 1+2=5. \\ 11111 & 5. & 5-3=2. \\ & & 5\div 3=1\ (2\text{ rem.}) \end{array}$$

By 4.

```
1 1 1 1    4.  4+1=5, 1+4=5.
        1  1.  4×1+1=5.
   ─────
   1 1 1 1 1  5.  5−4=1.
               5÷4=1 (1 rem.)
```

31.

3−1=?	1×3+1=?	5−2−2=?
5÷2=?	2×2+1=?	5×1=?
4−2=?	5−2=?	5÷5=?
2×2=?	3+1+1=?	5−4=?
4÷3=?	4−2−2=?	3×1+2=?
5−3=?	3÷3=?	4÷1=?
4+1=?	5÷4=?	2+2+1=?
5÷3=?	2+2=?	

32. 5 is 1 more than?
    2 is 1 less than?
    3 is 2 less than?
    4 is 2 more than?
    What number added to 2 will make 5?
    5 is 4 more than?
    3 is 1 more than?
    5 is 2 more than?
    2 is 2 less than?
    3 is 3 more than?
    1 is 2 less than?
    5 is how many times 1?

(Since 2 is added twice and 1 is added to the result to get 5, 2 times 2 + 1 = 5 and since 2 may be subtracted from 5 twice and 1 will remain, 2 is contained in 5 twice and 1 remainder, or 5 ÷ 2 = 2 [1 rem.].)

33. Teach pupils to write and read Roman notation to V. Teach pupils to count by 2's from 2 to 50, and back to 2, and from 1 to 11 and back.

34. For rapid solving.

5 − 2 − 3 + 2 × 2 − 3 + 2 = ?   Ans. 3.
2 × 2 + 1 − 3 × 1 × 2 − 3 + 3 = ?   Ans. 4.
4 − 1 + 2 − 3 × 2 − 1 − 2 × 3 − 1 = ?   Ans. 2.
3 + 2 − 1 − 2 × 2 − 3 − 1 = ?   Ans. 0.
5 − 3 + 2 − 3 × 3 − 1 + 2 = ?   Ans. 4.
2 + 2 − 3 × 2 + 3 − 2 + 1 − 2 = ?   Ans. 2.
2 + 1 − 2 × 3 − 3 + 2 × 2 is how much more than 1 ?   Ans. 3 more.
3 + 2 − 1 − 2 + 1 − 2 × 3 − 1 = ?   Ans. 2.

Review these frequently.

35. Review counting by 2's.

Teach to count by 2's, commencing with 1, to 21.

36. How many must I add to 3 to get 5? How many must be taken from 5 to get 3? Why? (Because 2 added to 3 makes 5.)

37. How many times 2 must I add to 1 to get 5?

I have taken away twice 1 from a certain number and 2 remains. What number was it?

I have taken 2 from a certain number and 1 remains. What number was it?

I have added 2 to a certain number and have 3. What number was it?

How many gallons are 2 quarts. Ans. None.

John had 5 dimes; he bought 2 copy books, each of which cost 2 dimes. How many dimes did he keep? (Illustrate, using dimes.)

George read a lesson once. Helen read it as many times as he did and two times more. How many times did she read it?

A father had 5 peaches and gave them to his 3 children; he gave the oldest 1 peach, and gave to each of the others an equal number; how many did each of the younger children receive?

A boy has 2 cents, he finds 2 cents; how many will he have to earn to have 5 cents?

1 is $\frac{1}{2}$ of what number?

James has 5 marbles, he loses 2; how many more than 2 has he left?

2 boys are passing my house, and each boy

is driving 2 goats ; how many goats are passing my house?

David rode 1 horse from the pasture to the barn and at the same time led 2 others ; how many horses did he bring to the barn?

Jane had 5 chickens. A rat ate 1 of them, and then a cat ate half of what were left and 1 more ; how many lived?

A boy, having 4 pockets, has 2 apples in 1 of them ; one pocket is empty, and he has 1 apple in each of the other pockets : how many apples has he?

38. Teach counting by 2's from 1 to 21 and back. Review former counting. (Do not teach all of this before giving other exercises, but require some of this kind of exercise daily until as much as is denoted above has been accomplished. These directions apply to future countings.)

SIX.

Schedule:

39. Measuring by 1. (Teach pupils to make these schedules.)

1 1 1 1 1 1. 6.

```
  1 1   1+1+1+1+1+1=6.
  1 1   1×6=6.
  1 1   6−1−1−1−1−1=1.
  1 1   6÷1=6.
  1 1
  1 1
  ───
  6 6
```

By 2.
```
           1 1 2   2+2+2=6.
           1 1 2   2×3=6.
           1 1 2   6−2−2=2, 6−2−2−2=0.
  ─────────────
  1 1 1 1 1 1 6   6÷2=3.
```

By 3.
```
           1 1 1 3   3+3=6.
           1 1 1 3   3×2=6, 2×3=6.
  ─────────────
  1 1 1 1 1 1 6    6−3=3, 6−3−3=0.
                   6÷3=2.
```

By 4.
```
        1 1 1 1  4.   4+2=6.
              1 1  2.   4×1+2=6.
  ─────────────
  1 1 1 1 1 1  6.   6−4=2, 6−2=4.
                    6÷4=1 (2 rem.)
```

By 5.
```
        1 1 1 1 1  5.   5+1=6, 1+5=6.
                1   1.   5×1+1=6.
  ─────────────
  1 1 1 1 1 1  6.   6−5=1, 6−1=5, 6−
                    5−1=0.
                    6÷5=1 (1 rem.)
```

FIRST STEPS AMONG FIGURES.   29

40.

$5-2=?$ $2\times 2+1=?$ $4-2=?$
$3+3=?$ $1+1+4=?$ $1+2+1=?$
$2\times 2=?$ $6\div 2=?$ $6-2-2-2=?$
$1+4+1?$ $3+2=?$ $4\div 1=?$
$5\div 1=?$ $5-1-1=?$ $2\times 2+2=?$
$3\times 1=?$ $6-2=?$ $5-3=?$
$6-2-2=?$ $1\times 5=?$ $4-1-1-1=?$

41. 6 is 3 more than?
   What is ½ of 6?
   6 is 4 + how many?
   6 is 2 times what number?
   How many times can you take 4 from 6?
   6 is 3 times what number?
   4 less than 6 is?
   6 is 2 + ?
   What number is 3 less than 6?
   What number is half of 4?
   6 is 1 + ?

42. For rapid solving.
   $1+2+2=?$ $3+2-1+2=?$
   $1+1+1+2=?$ $2+2+2=?$
   $3+2-1\div 2\times 3=?$
   $5-2+3-2-3+1\times 2+1-3+1?$ Ans. 3.
   $3+1-2\times 3-3+1-2+1\times 2-2\div 2=?$
      Ans. 2.

$2+3-2\times 2-5\times 3+2-3\times 1+2$? Ans. 4.
$4-3+2-1\times 2+2+2+2+2+2+2$? Ans. 16.
$5+2+2+2+2+2+2+2+2+2+2+2$? Ans. 27.

43. Count by 3's from 3 to 12. Review counting.

Count by 3's from 3 to 18. Review the counting already taught. Teach Roman notation to X.

44. For rapid solving.

$6-2-2+3-2\times 2+2+2$? Ans. 10.
$2\times 3-4+3-1+2+2+2+2+2+2+2$? Ans. 18.
$5-3+1+3-2\div 2+3+2+2+2+2+2+2+2$? Ans. 19.
$1+3-2+3+1\div 3+2+2+2+2+2=$? Ans. 12.
$5-2-2\times 2-1+4-3\times 3+3+3+2+2+2=$? Ans. 18.
$2+3-2+3\div 2+2+2+2+2+2+2+2=$? Ans. 17.
$4-3+2+2-3\times 2+2+2+$? Ans. 8.
$6\div 3+3-2\times 2+3+3+3+2+2+2+2$? Ans. 23.

45. Examples in addition may be written on

the board, a few for the pupils to solve each day on their slates and bring to class; and additional examples may be read to the class and solved on the slates or on the board.*

46. 3 cts. is 2 cts. more than John has; how many has he?

3 cts. is 1-2 of what money Jane has; how much has she?

2 cts. is 3 cts. less than an orange cost; how many cts. did it cost?

1 is how many less than 6?

6 cts. will buy how many 3 cent. stamps? 1 ct. stamps? 2 ct. stamps? 5 ct. stamps?

George had 1-2 of 6 cents; how many had he?

Clara had 6 flowers; she gave them to her father and mother. If she gave each of them the same number, how many did her father get? Her mother?

Charles had 6 cts.; he lost 2 of them; how many had he left?

Carrie had 6 peaches: she gave her father 2 of them and her mother 2; how many did she keep?

3 cts. is 3 cts. more than Byron's money; how much money has he?

* See T. Ed., p. 84, and P. Ed., p. 33.

47. $3+2-1+2-5+3=?$ Ans. 4.

1-2 of $6+2-3$ is what part of 4? Ans. 1-2.

$1+2+2-3-1+3-2=?$ Ans. 2.

$3+2-1-3+1\times3+3+3+3+2+2=?$
Ans. 19.

$6-4+2-1\times2-2+2+3+3+2+2+2?$ Ans. 18.

I put down a number once and again and again to get 6. What is the number?

48. Count by 3's from 3 to 24. Review the counting.

From what number can you take 2×2 and keep 1?

What number must I double to get 4?

What number is one less than 5?

What is 1-2 of 4?

What is ½ of the number 1 less than 5?

What number is one less than ½ of 4?

Augusta had 5 cents. She lost 1 of them, and spent ½ of what she had left, and then found 3 cents. How many had she then? (Solve one step at a time.)

49. For the method of making series like the following, see preface. For method of use see pages 42 and 43.

For addition.*

---

* Teach each series thoroughly before taking the next.

a	b	c	d	e
4 2 0	1 3 4	2 0	1 3	
1 2 1	2 1 2	1 2	1 2	

For subtraction.

a	b	c	d	e	f	g
5 2	5 3	6 2	5 4	3 1	6 3 4	
3 1	2 1	2 2	1 2	2 1	3 3 1	

For multiplication.

a	b	c	d	e	f	g
2 0	3 1	2 0	3 1	2 0	3 1	
1 2	0 1	2 0	1 2	0 1	2 0	

For division.

a	b	c	d	e	f
6 2	3 0	2 3	0 6	1 4 0	
3 2	1 2	1 3	1 2	1 2 3	

For addition. (Re-arranged.)

a	b	c	d	e	f	g	h
4 3	2 0	2 1	3 0	4 2	0 2	1 3	
3	2 1	3 2	1 3	2 1	2 3	1 2	

For subtraction. (Re-arranged.)

a	b	c	d	e	f	g	h
4 1	5 6	3 4	3 3	2 5	2 5	6 5	
2 1	3 2	3 1	2 1	2 3	1 2	3 1	

50. Teach Roman notation to XVIII.

Review counting by 2's commencing with 2, and also with 1, to 60 and 61.

Teach counting by 3's commencing with 3 to 30.

The examples given under previous numbers should be frequently reviewed, so that the pupil may become quite familiar with the formation and use of numbers.

SEVEN.

51. At this stage pupils should make the schedule from their memory on the plan of those already given.

Schedule:

Measuring by 1.
    1 1 1 1 1 1 1 7.
    $1+1+1+1+1+1+1=7$.
    $1 \times 7. \quad 7$.
    $7-1-1-1-1-1-1=1$, or $7-1-1-1$
       $1-1-1-1=0$.
    $7 \div 1 = 7$.

By 2.
    1 1 2. $2+2+2+1=7$.
    1 1 2. $2 \times 3 + 1 = 7$.
    1 1 2. $7-2-2-2=1$.
      1 1.

    7 7. $7 \div 2 = 3$ (1 rem.)

Teach pupils to write Arabic to 100.

By 3.
    I I I 3. $3+3+1=7.$
    I I I 3. $3\times 2+1=7.$
        I I. $7-3-3=1.$
_____
I I I I I I I 7.
           $7\div 3=2$ (1 rem.)

By 4.
    I I I I 4. $4+3=7.$
    I I I 3. $4\times 1+3=7.$
_____
I I I I I I I 7. $7-4=3.$
           $7\div 4=1$ (3 rem.)

By 5.
    I I I I I 5. $5+2=7.$
        I I 2. $5\times 1+2=7.$
_____
I I I I I I I 7. $7-5=2.$
           $7\div 5=1$ (2 rem.)

By 6.
    I I I I I I 6. $6+1=7.$
          I I. $6\times 1+1=7.$
_____
I I I I I I I 7. $7-6=1.$
           $7\div 6=1$ (1 rem.)

52.
$5+1=?$ $7-5=?$ $6\div 4=?$
$6-2=?$ $1\times 5+1=?$ $7-3=?$

$5 \div 1 = ?$ $3 \times 2 + 1 = ?$ $2 \times 2 + 2 = ?$
$2 \times 2 = ?$ $6 - 2 - 2 = ?$ $7 - 2 - 2 - 2 = ?$
$3 + 2 = ?$ $2 \times 2 = ?$ $1 \times 5 + 2 = ?$
$4 - 2 = ?$ $7 \div 4 = ?$ $5 - 1 = ?$
$6 + 1 = ?$ $3 \times 2 = ?$ $7 \div 3 = ?$
$2 \times 2 + 1 = ?$ $2 \times 2 + 3 = ?$

53. 7 is 3 more than?
What is ½ of the number 1 less than 7?
7 is 1 more than twice what number?
What number is 3 less than 7?
What number must be added to 3 to get 7?
How many times can you subtract 2 from 7?
7 is 4 + ?   3 is ½ of?
I added 3 to a certain number and got 5; what number was it?
1 pear is what part of 7 pears?
2 pears are how many times 1 pear?
How much must be taken from 7 to leave 3?
What number must be added to 2 to get 7?
How many times 3 must I add to 1 to get 7?

Count by 3's from 3 to 42. Review the counting.

How many times can you subtract 5 from 7?

54. 6 is double what number? 3 is 2 less than?
What number is one less than 7?
What is ⅓ of the number 1 less than 7?

55. For addition.

a	b	c	d	e	f	g
314	203	120	312	0131	420	
123	412	312	342	3441	234	

For subtraction.

a	b	c	d	e	f	g
456	277	615	454	353	2643	
432	143	412	312	341	2312	

For multiplication.

a	b	c	d	e	f
31	03	102	10	22	
12	32	132	12	31	

For division.

a	b	c	d	e	f
26	01	40	36	023	
13	21	21	32	321	

Same re-arranged. *

For addition.

a	b	c	d	e	f	g
241	302	413	021	302	4130	
123	412	341	231	234	1234	

For subtraction.

a	b	c	d	e	f	g
435	256	534	634	741	6375	
432	143	412	312	341	2312	

For multiplication.

a	b	c	d	e	f
20	12	03	12	031	
12	12	31	23	123	

* For review.

For division.

a	b	c	d	e	f
32	60	40	63	201	
12	31	23	23	121	

56. After earning 3 cts., Fanny had 5 cts.; how many had she before?

How many must you add to 3 to make 7?

Nellie has 2 pencils, and Sarah has 1 more than Nellie; how many have *both* of them?

Marcus has 4 marbles and Arthur has 2 less; how many have both?

Mary has 6 pins and Stella has 2; how many more has Mary than Stella?

57. When the pupils become listless or restless, or a minute or two of spare time is at command, the following examples and like examples given through the book will be found both useful and interesting. Use them often.

$3+2+2-3+2 \div 2-1 \times 2+3-2-3+2$?
Ans. 4.

$5-3+1 \times 2 \div 3+2-3+4-2 \times 2-4 \times 3+1$? Ans. 7.

$4+3-2-3+1 \times 2-3+2-3 \times 3 \div 2+4-2$? Ans. 5.

$7-2-3+4-2+3-5+1 \times 2-3+3-4-2$? Ans. 0.

$2+5-3\div2-2+3\times2-2\div2\times3\div2-1$?
Ans. 2.
$5-2-2\times7-5\times2-3+1\times3-4+3-2+4-4$? Ans. 3.
$6-4+3+2-1\div3\times2+3-4+2+2+3+3+3$? Ans. 16.

59. Teach the pupils to read numbers to 1001. Show them that the figure in the third place represents hundreds (or is named-hundreds.)

Read the following numbers:

1. 432.	2. 395.	3. 216.	4 743.
5. 341.	6. 704.	7. 435.	8 914.
9. 370.	10. 891.	11. 706.	12. 289.
13. 514.	14. 780.	15. 981.	16. 709.

Teach the pupils that the fourth place represents thousands (or is named thousands.) Teach the pupils to write a comma between hundreds and thousands before reading a number.

Read:

17. 1395.	18. 3741.	19. 5416.	20. 7308.
21, 9150.	22. 7075.	21. 4118.	24. 9400.
25. 7804.	26. 1000.	27. 7050.	28. 8004.

60. Teach pupils to write numbers to 1,000 in Arabic, and in Roman to XX.

1. Write two hundred forty-five (in **Arabic.**)
2. " five hundred sixty one.
3. " one hundred thirty four.
4. " seven hundred twenty-one.
5. " three hundred eighty-six.
6. " four hundred sixteen.
7. " nine hundred twenty-one.
8. " six hundred thirty-two.
9. " eight hundred seventy-nine.

61. Give more examples like the above until the pupils write them readily. Then give the following:

10. Write three hundred nine.
11. " eight hundred forty.
12. " four hundred fifteen.
13. " six hundred thirty-seven.
14. " two hundred ninety.
15. " sixty-four.
16. " seven hundred two.
17. " five hundred six.
18. " three hundred thirty.
19. " four hundred one.

Review these often.

62.

$3+2+3+3+2+2+3+3+3+3+2+2?$
Ans. 31.

$4+2 \div 3+3+3+3+3+3+2+2+2+2+2$? Ans. 27.

$2+3+3+2+2+3+3+3+3+2+2$? Ans. 28.

$4-2 \times 3-4+1 \times 2-3+2+1 \div 3 \times 3+2+2+3$? Ans. 13.

$6 \div 2-1 \times 3-4+1 \times 2+3+3+3+2+2+3+3$? Ans. 25.

$7-4 \times 2-2+3+3+3+2+2+2+2+3+3+2$? Ans. 29.

$5+3+3+3+2+2+2+2+3+3+3+2$? Ans. 33.

$7-5 \times 3-2 \div 2+3+2+2+3+3+3+3+2+2$? Ans. 25.

Some boys are sliding down hill. There are 3 sleds and two boys on each sled, how many boys are there?

James had 7 apples; he ate one and gave his sister half of the rest. How many did he give his sister?

John had two apples; he cut each of them in halves. How many halves had he?

How many horses in 2 two-horse teams?

A stingy boy had 5 sticks of candy; he would neither eat any nor give any away. How many did he keep?

A generous boy had 3 sticks of candy; he gave his sister 2 sticks and he ate half a stick. How many had he left?

Ralph had 5 peaches; he gave 2 of them to his little sister, 1 to his father, 1 to his mother, and ate 1 himself. How many had he left?

64. Count by 3's from 3 to 50.

Count by 2's from 2 to 60.

Count by 2's from 1 to 61.

Teach Roman notation to XXVIII.

Write in letters 19, 13, 21, 14, 11, 8, 16, 25, 12, 26, 17, 9, 25, 27, 18.

66. Count by 3's from 1 to 22.

Review counting.

EIGHT.

67. Pupils may make the schedule like previous ones.

68. Give only a small part of these series each day and give with it slate examples from pp. 93 to 94, and oral exercises in the examples following the series. By this variety much mo.e work may be accomplished without the weariness resulting from too much sameness.

The series should on each succeeding day be

FIRST STEPS AMONG FIGURES. 43

reviewed.* For a review after completion take the re-arranged series.

For addition.

a	b	c	d	e	f	g
3 5 2	4 1 3	5 2 4	1 3 2	4 1 3	5	2 4 1
1 2 3	4 1 2	3 4 1	2 3 1	2 3 4	1	2 3 4

For subtraction.

a	b	c	d	e	f	g
6 5 3	6 3 6	8 5 2	8 5 7	4 7 6	4 7 5	4
4 1 2	3 1 2	3 2 1	4 3 2	3 4 1	2 3 4	1

For multiplication.

a	b	c	d	e
4 2	1 4	2 3	1 3	2 1
1 2	1 2	3 1	2 2	4 3

For division.

a	b	c	d
4 3 4	1 8 2	6 3 2	6 3 8
1 3 2	1 2 1	3 1 2	2 3 4

For addition. Re-arranged.

a	b	c	d	e	f	g
1 4 2	3 1 4	2 5 3	1 4 2	5 3 1	4 2 5	3
1 2 3	1 2 3	4 1 2	3 4 1	2 3 4	1 2 3	4

For subtraction.

a	b	c	d	e	f	g
3 4 5	6 2 7	6 6 5	4 8 4	8 7 5	5 6 7	3
2 1 3	2 1 3	4 1 2	3 4 2	3 4 1	4 3 2	1

* Leave on the board the previous day's lesson in series and add to it as much more of the series as can be mastered with the review.

69. Teach Roman notation to XXX.
What 2 equal numbers make 8?
What is half of 8?
What number is one less than half of 8?
What number can you double and get 8?
From what number can you take 2 × 3 and have 1 left?
Count by 3's from 1 to 40.
I write a number four times and add. I get 8, what is the number?
Henry had half of 8 cents. How many had he?
8 cents will buy how many 1-cent stamps?
4-cent stamps (I. Revenue)? 2-cent stamps? 5-cent stamps? 3-cent stamps?
3 lemons is ½ of how many lemons?
Lewis brought 6 eggs from the barn; he broke half of them. How many whole ones were left?
George has 3 cents, he finds 2 cents; how many must he earn to have 8 cents?
William had 2 sticks of candy, he ate half of a stick, and his sister half a stick; how much candy had he left?
What number is 1 less than half of 6?
Jane's bird hatched 3 young birds and there

were 2 eggs which did not hatch; how many eggs in the nest at first?

What is half of the number 1 less than 7?

William bought 3 marbles at 2 cents each; how much should he pay for them?

How many tops at 3 cents apiece can Edward buy for 8 cents? For 7 cents?

6 cents is 4 cents more than Robert's money, how much money has he?

$3 \times 2 - 4 + 3 - 2 \times 1 - 1 \times 3 + 1 = ?$ Ans. 7.

$7 - 3 + 2 \div 2 - 1 \times 4 - 3$ is how many less than 8? Ans. 3.

$2 + 3 - 1 \div 2 - 1 \times 6 \div 2 + 5 - 2 \div 3$ is ½ of what number? Ans. 4.

$6 + 2 - 5 + 1 \div 2 + 1 \times 2$ is 1 less than? Ans. 7.

$5 + 2 - 1 \div 3 \times 4 \div 2 + 3 + 3 + 3 + 3 + 3 + 2 + 2$? Ans. 23.

$3 + 3 \div 3 - 1 \times 3 + 3 + 3 + 3 + 3 + 3 + 2 + 2 + 3$? Ans. 25.

$7 - 2 - 1 \div 2 \times 4 + 2 + 1 + 3 + 3 + 2 + 2 + 2 + 2 + 3 + 3$? Ans. 31.

$5 + 3 \div 4 \times 3 + 3 + 3 + 3 + 3 + 2 + 2 + 3 + 1 + 3 + 2$? Ans. 31.

$8 \div 2 - 1 \times 2 \div 3 + 3 + 3 + 3 + 2 + 2 + 3 + 3 + 3 + 2 + 3$? Ans. 29.

$3+2+2+3+3+2+2+2+3+3+2$? Ans. 27.

$4+3+2+2+3+3+3+2+2+3+3+2+2$? Ans. 34.

$5+3+2+3+3+2+3+3+2+2+2+3$? Ans. 33.

$4+2+3+3+3+2+2+3+3+2+2$? Ans. 29.

$7+3+3+3+3+2+2+3+3+3+2+2$? Ans. 36.

$4+2+3+3+3+2+2+2+2+3+3+3+3+3$? Ans. 38.

$5+3\div2+2\div3\times2+3+3+3+3+3+2+2+2+2$? Ans. 27.

Oral exercise. Place the figures of any of these examples which involve only addition in a column on the board, let one pupil add them upward, then another downward, then another add them upward, but commence with 1 and so get a result 1 greater than before, add downward in the same way commencing with one; then use 2 instead of 1 and also 3 instead of 2. This makes 8 different examples instead of one and gives excellent practice.

How many feet have 2 pigs?

8 is 2 more than what number?

8 is 2 more than twice what number?
How many legs have 3 hens?
What number is 5 less than 8?
How many times can you substract 3 from 8?
How many legs have a rabbit and a bird together?
8 peaches are how many times 2 peaches?
1 pencil is what part of 8 pencils?
How many wheels have 4 sulkies?
What number must I add to 5 to get 8?
4 is half of what number?
From what number can you take 2 times 2 and have 3 left?
2 is how many less than 6?
Count by 2's from 1 to 62.
" " 3's " 3 to 60.
" " 2's " 2 to 60.
" " 3's " 1 to 61.
Teach Roman notation to XXXIX.
Write in letters 26, 18, 34, 9, 16, 22, 37, 19, 35, 21, 36, 39, 17.
7 is how many more than 5?
John had 7 sticks of candy; he ate 3 and gave away ½ of the rest. How many did he give away? How many did he keep?
Count by 3's from 2 to 62.

48   FIRST STEPS AMONG FIGURES.

Teach Roman notation to L.

Write in letters 34, 16, 25, 19, 30, 28, 34, 17, 26, 33, 46, 37, 22, 50, 44, 27, 33, 49.

Write in figures XVIII, XXXIV, XV, XL, XXVI, XII, XXXVI, XIV, XVII, XIX.

NINE.

Pupils make a schedule.
For addition.

a	b	c	d	e	f	g
415	324	153	624	136	241	362
123	412	345	123	412	345	234

h	i	j
451	532	
511	235	

For subtraction.

a	b	c	d	e	f	g
736	498	756	948	585	384	757
412	345	125	312	341	232	345

h	i	k
672	69	
321	15	

For multiplication.

a	b	c	d	e	f	g	h	i
24	10	21	30	21	30	24	130	
12	31	24	12	31	23	42	234	

FIRST STEPS AMONG FIGURES.

For division.

a	b	c	d	e	f	g	h	i
30	44	30	61	84	94	42	062	
31	24	12	31	22	32	12	321	

Re-arranged and including o.

For addition.

a	b	c	d	e	f	g	h
526	301	452	831	740	523	214	365
123	435	123	051	226	341	723	614

i	k	l
236	14	
520	34	

For subtraction.

a	b	c	d	e	f	g	h
268	643	485	756	796	467	967	759
123	412	453	215	430	213	465	123

i	k	l
835	6839	
404	3215	

For multiplication.

a	b	c	d	e	f	g	h	i
31	20	31	20	31	24	14	20	
12	42	23	12	34	22	11	14	

For division.

a	b	c	d	e	f	g	h
32	09	40	61	86	04	82	343
31	23	41	21	23	41	42	121

How many 1's make 9?

From what number can I take 9 1's and have nothing left?

How many 3 cent stamps can you buy for 9 cents? 2 cent stamps? 1 cent? 5 cent? 4 cent?

How many oranges at 9 cts. each can you buy for 9 cts.?

How many times can I take 2 from 9 and have 1 left?

George had 9 peaches; he ate one and gave you half of what were left, how many did he give you?

What number taken 3 times will make 9?

Henry gave each of his 3 playmates 3 plums. How many did he give away?

What is half of the number 1 less than 7?

William has 9 wheels. He has wheels for how many three-wheeled velocipedes?

5 and how many make 9?

What number taken twice and 3 added makes 9?

What is half the number 1 less than 9?

Theodore had 9 marbles, he lost 4 of them, how many had he left?

4+3+?=9.

8 is 3 and how many?
1-3 of 6 and how many make 6?
Count by 4's from 4 to 20. Review.
What number is 1 less than ½ of 8?
9=6+?
5 from 9 leaves?
What cost 4 lemons at 2 cts. each?
What cost 2 marbles at 3 cts. each?

$5+3-4-2+3+4-6=?$ Ans. 3.
$3+4-2+4\div3+3\div2+1=?$ Ans. 4.
$6+3-5\div4-1\times2+7+2=?$ Ans. 9.
$2\times3+2-5+1\div2+3-4+2+5=?$ Ans. 8.
$4-3+5\div3+7\div3\times2+1-4=?$ Ans. 3.
$6+2-6\times4-5\times3+3+3+2+3?$ Ans. 20.
$3+5\div4+4\div2+2+2+3+3+2+2+3+3+2?$ Ans. 25.
$2\times3-3\times3-1\div2+2+2+1+3+3+3+3+2+2-4?$ Ans. 21.
$5+3+3+2+3+2+2+3+3+3+2+3+2+3?$ Ans. 41.
$8+3+3+3+2+2+3+3+2+3+2+2+2+3?$ Ans. 39.
$7+2+3+3+2+2+2+3+3+3+2+3+2+3+2?$ Ans. 42.

3+2+2+3+2+3+3+2+ 2 +3 + 3+2
+2? Ans. 32.
5+2+5+3+2+2+ 3 +3+3+3 +2 +3
+2? Ans. 38.
8+3+2+2+3+2+ 3 + 3 +2+2+3+3
3+3? Ans. 42.
4+3+2+3+3+ 3 +2+2+ 3 +2+3+3
2+2? Ans. 37.

Review these often.

Count by 2's, commencing with 2 and with 1, to 60 and 61.

Count by 3's, commencing with 3, 2 and 1, to 51, 53 and 52.

Count by 4's, commencing with 4, to 32 and back.

James having 9 apples ate 1, and gave the rest to his sisters, giving them 2 each; how many sisters had he?

George had 9 oranges; he ate one of them; if he were to give you half of what were left, how many would you get?

Charles having 9 pears sold 3, and gave his sister half of what he had left; how many did he give his sister?

Teach Roman notation to LXX.

Write in letters 67, 44, 36, 59, 62, 46, 28, 16.

FIRST STEPS AMONG FIGURES. 53

Write in figures XVII, LXV, LXVII, LXIV, LXIX, XXIV, XIX.

Pupils read the following, which should be copied on the black board, and the pupil who reads any number to point off the periods himself before reading.

1. 6305. 2. 7020. 3. 8005. 4. 9400. 5. 1641.
6. 6780. 7. 5416. 8. 8605. 9. 5400. 10. 7508.
11. 4870. 12. 5718. 13. 5851. 14. 6504. 15. 5790.
19. 1432. 17. 9007. 18. 5000. 19. 7400. 20. 8040.
21. 4637. 22. 5819. 23. 7990. 24. 7803. 25. 7001.

Teach Arabic notation to 10,000.

Teach the pupils carefully, as being of the utmost importance, that they should place a comma after the number expressing thousands and before they write the units period. Teach them that units period takes three places, and show them that when the number does not fill the three places, the places on the *left* must be filled with ciphers.

The following are to be read by the teacher and the pupil is to write them in Arabic. Review these often.

1. 4,573 2. 3,240 3. 5,296 4. 7,315
5. 2,324 6. 7,560 7. 1,427 8. 3,670 9. 7,305
10. 5,741 11. 2.816 12. 5,980 13. 6,407 14. 4,300

15. 9,706 16. 4,315 17. 8,590 18. 1,731
19. 7,800 20. 5.004 21. 3.060 22. 5,104
28. 6,003 24. 8,600 25. 9,419 26. 8,040
27. 5,900 28. 4,307 29. 5,009 30. 9,016
31. 9,360 32. 9,070 33. 7,049 34. 3,900
35. 7,008 36. 7,080 37. 5,700 38. 7,000
39. 5,875 40. 3,716 41. 9,060 32. 5,800
43. 6,904 44. 6,008 45. 7.600 46. 8,009

Cut an apple into 3 equal pieces and teach the pupils that we call one piece one third. Break a stick of candy into 3 equal pieces and so illustrate the same thing.

In the same way teach one-fourth by 4 divisions instead of 3 ; then one-fifth, one-sixth, &c. When the pupils are familiar with this show them that of any 3 equal things, one of those things is one-third ; of 4 equal things, one of them is one-fourth. Do not leave this subject until the pupils are very ready with their answers to the following questions :

An apple is cut in 5 pieces; what do we call 1 piece ?

An apple is cut in 3 pieces ; what do we call 1 piece ?

An apple is cut in 6 pieces ; what do we call 1 piece ?

An apple is cut in 10 pieces; what do we call 1 piece?

An apple is cut in 4 pieces; what do we call 1 piece?

One apple is what part of 7 apples?

One apple is what part of 9 apples?

One apple is what part of 6 apples?

One orange is what part of 4 oranges?

One pencil is what part of 8 pencils?

A boy having 5 apples gave away one of them; what part of his apples did he give away?

A little girl had 6 peaches; she gave one-third of them to her brother. How many did she give him?

(Teach the pupils that they get one-third of a number by dividing by 3; ¼ of a number by dividing by 4, &c.)

A boy gave away one-fourth of his 4 marbles; how many did he give away? How many did he keep?

If I divide 6 apples equally among 3 boys, what *part* of them do I give each boy? How *many* do I give each boy?

TEN.

Pupils make schedule.

FIRST STEPS AMONG FIGURES.

For addition.

a	b	c	d	e	f	g	h	i
536	247	503	247	535	245	623	043	6234
234	5^2	364	623	450	235	236	742	3405

For subtraction.

a	b	c	d	e	f	g	h	i
1098	758	676	6107	1074	389	1085	983	5969
623	406	223	6	45	532	054	323	643	2345

For multiplication.

a	b	c	d	e	f	g
352	313	240	132	421	521	20
123	012	415	231	203	124	53

For division.

a	b	c	d	e	f
842	920	616	1038	035	44
412	313	213	212	231	24

Re-arranged.

For addition.

a	b	c	d	e	f	g	h	i
635	264	635	207	463	257	430	524	3524
234	502	345	652	345	203	464	235	2346

For subtraction.

a	b	c	d	e	f	g	h	i
8410	839	796	869	5710	897	759	7105	86510
52	3	406	543	223	54	5	623	235	0	42	345	6

For multiplication.

a	b	c	d	e	f	g	h	i
42	14	24	31	25	02	31	25	31
12	12	30	12	.41	35	23	12	31

For division.

a	b	c	d	e	f	g	h	i
18	60	32	85	44	02	106	310	94
12	34	12	41	12	31	22	3 5	34

What is ½ of 10 buttons?

Lucy had 9 pins; she lost 4 of them and then found 1; how many had she then?

4 and how many make 10?

10 boys were out in a sail boat; 1 more than half of them were drowned; how many were drowned?

Minnie had 9 cents; she spent 5 of them and lost 4; how many had she left?

10 cents will buy how many 3 cent stamps?

How many 2 cent sticks of candy can you buy for 10 cents?

How many cents will 4 two-cent marbles cost?

Lucy had 5 cents and her mother gave her 4 cents, how many cents had she then?

Show the pupils that we call 2 and 2 equal numbers, that they are equal to each other, also 1 and 1 are equal numbers, 3 and 3, &c.; 2 and 1, or 2 and 3, or 1 and 3 are unequal numbers.

What 2 equal numbers make 4?

What 2 unequal numbers make 4?

What equal numbers make 3?
What unequal numbers make 3?
What 2 equal and 1 unequal numbers make 5? (Two answers.)
How many wheels have 5 sulkeys?
How many wheels have 3 three-wheeled velocipedes?
Mary was bringing in 10 eggs in her apron, she broke 2 less than half of them; how many did she break?
How many were unbroken?
John had a string 10 yards long and William had one three yards long; how much longer was John's string than William's.
6 and how many make 10?
What number taken from 10 leaves 7?
What cost 5 two-cent stamps?
Henry had 10 miles to walk, he has walked 4 of them; how much farther has he to walk?
10 beans are how many times 2 beans?
8 boys were playing "snap the whip," 6 of them kept hold of hands; how many were there that did not let go?
What number added to 3 will make 10?
Count by 4's, commencing with 4, to 60.

FIRST STEPS AMONG FIGURES.

Review counting by 2's and 3's.
For rapid solving.

$4+3-1 \div 2 \times 3+1 \div 5 \times 4+4 \div 2+3+3$
$+3+3+2+2+2 = ?$ Ans. 24.

$7+2-3+4 \div 2-2 \times 2+3+3+3+3+2$
$+2+3+3+4+4+4 = ?$ Ans. 40.

$4+3+3+3+3+3+2+2+1+4+4+4$
$+4+4 = ?$ Ans. 44.

$3+1+4+4+4+4+4+3+3+3+2+2$
$+3+3+2+3 = ?$ Ans. 48.

$5+5 \div 2+3 \div 2-2 \times 3+2+4+4+4+3$
$+3+3+3+2+2 = ?$ Ans. 36.

$7+3+3+3+4+4+4+4+3+2+2+3$
$+3+3+3 = ?$ Ans. 51.

$5+3 \div 4+4 \div 2+3+3+3+2+2+3+3$
$+3+3+2+2+3+3 = ?$ Ans. 38.

$4+3+3+3+3+3+3+2+4+4+4+4$
$+3+2+2+2+3+3 = ?$ Ans. 55.

$10 \div 5 \times 3-2 \times 2 \div 2-2 \times 3+3+2+2+$
$3+1+4+4+4 = ?$ Ans. 29.

$5+3+4+4+4+4+4+4+4+4+3+3$
$+3+2+2+2+2 = ?$ Ans. 57.

$7+2+2+3+2+4+4+4+3+3+3+2$
$+2+3+4+4+3 = ?$ Ans. 55

$5+3+4+4+4+4+3+2+3+4+8+8$
$+4+8+2+8 = ?$ Ans. 74.

$4+3+2+1+4+4+4+4+3+2+3+3$
$+2+3+3+3+3= ?$ Ans. 51.

For more practice see pp. 51 and 45.

Count by 4's from 2 to 34.
"   " 4's " 4 " 60.
"   " 3's " 3 " 60.
"   " 3's " 1 " 61.
"   " 3's " 2 " 62.
"   " 2's " 2 " 60.
"   " 2's " 1 " 60.

Write in letters 36. 41, 16, 64. 56. 47. 69.

Write in figures XLV, XV, LXIV, LXXXV, XXVII.

What equal numbers will make 6? (3 ans.)

What unequal numbers will make 6? (Several answers.)

What 2 equal numbers and 1 unequal number will make 6?

The following numbers are to be copied on the blackboard and the pupils are to be required to point them off in periods and read them.

1. 91017. 2. 86700. 3. 90007. 4. 14071.
5. 70000. 6. 50010. 7. 38419. 8. 74058.
9. 60800. 10. 16040. 11. 3000. 12. 7014.
13. 10061. 14. 3020. 15. 7003. 16. 8500.
17. 17500. 18. 3540. 19. 67374. 20. 86000.

What 2 equal numbers and 1 unequal number will make 7? (3 answers.)

What 3 equal numbers and 1 unequal number will make 7? (2 answers.)

What is 1-3 of 9?

1 is what part of 7?

Teach Arabic notation to 99,000.

Show the pupils that they should place a comma after the figures that express thousands before writing the unit period, and a period at the end of the number.

The following numbers are to be read by the teacher and written upon the blackboard or slates by the pupils.

1. 7,300. 2. 5,006. 3. 2,050. 4. 10,091.
5. 8,016. 6. 4,000. 7. 12,090. 8. 50,700.
9. 65,078. 10. 45,913. 11. 80,000. 12. 80,010.
13. 15,061. 14. 40,002. 15. 79,500. 16. 81,018.
17. 30,600. 18. 60,060. 19. 90,004. 20. 8,050.
21. 75,000. 22. 74,695. 23. 31,280. 24. 13,300.
25. 14,041. 26. 10,010.

Teach Roman notation to C.

Write in Roman 1. 64. 2. 49. 3. 97. 4. 76.

Write in Arabic 5. XIX. 6. LXXXIV. 7. XLI. 8. XXVII. 9. XVIII. 10. XXIX.

Write in Roman 11. 17. 12. 56. 13. 83. 14. 49.

Write in Arabic, 15. LXVIII. 16. XCII.

If the teacher prefers it, the pupils can buy their books (the Pupils' Edition) at this stage, and do more slate work than the teacher could have time to dictate to them, or copy upon the board for the pupils to copy and solve. If the pupils do not have their books the teacher will assign daily lessons from page 1, 2, &c., of the Pupils' Edition, doing it in connection with this work and thus carrying on that work together with the following work.

The teacher should now begin to give the parallel work of the Pupils' Edition in connection with that of the Teachers' Edition. The parallel pages are denoted by the numbers at the bottom of the pages in each.

PART II.

Counting two or more numbers into one number is called Adding, or Addition.

The number obtained by counting two or more numbers into one number is called the *sum* of those numbers.

For addition and multiplication. (5 and rev.)

a	b	c	d	e	f	g
352	413	524	135	241	352	4 1
234	523	452	345	234	523	4 5

For subtraction.

a	b	c	d	e	f	g
7410	746	758	693	696	857	8 5
32 5	432	523	452	345	432	5 4

For division.

a	b	c	d	e
41510	6165	20 92	20815	610 8
4 5 2	3 45	4 32	54 3	2 5 2

f	g
31225	4 12
3 4 5	2 3

See Pupils' Edition, p. 5.

It is well to require pupils to bring a written analysis of an example to recitation and to give the solutions of other examples orally in class in the same form, but there should be a large number of examples given, of which *only the answer* is to be given and that as soon as possible after the reading.

1. Susan had 4 cents and her mother gave her 3 more ; how many had she then ?

Solution. She had the sum of 4 cents and 3 cents, or 7 cents.

2. John has 5 marbles and James has 4 marbles ; how many have both ?

3. Lulu has 3 eggs in one hand and 2 in the other ; how many has she in both ?

4. Walter bought some candy for 4 cents, and some raisins for 5 cents ; how many cents did he spend?

5. Martha read 4 pages in the forenoon and 2 in the afternoon ; how many did she read that day ?

6. A boy had 3 pencils in one pocket and 5 in another ; how many had he in both ?

7. If a top cost 4 cents and a marble cost 2 cents, how many cents must a boy have to buy a top and a marble ?

FIRST STEPS AMONG FIGURES. 65

8. Jane bought 2 books; she had 3 before. How many has she now?

9. Henry walked 4 miles before dinner, and 4 after dinner; how far did he walk that day?

10. There are 3 barrels of apples under one tree, and two under another; how many under both?

Ask the pupils to bring examples of their own to recitation different from those given them. The teacher also will make additional examples, using pleasant facts about the school room, or the pupils, or their homes, something they have seen.

Count by 4's from 2 to 62.

Taking one number from another number is called subtracting, or Subtraction.

The number obtained by taking one number from another number is called the Remainder or Difference.

1. Joseph had 8 cents; he spent 5 cents for an orange. How many cents had he left?

Solution: He had left the difference between 8 cents and 5 cents, or 3 cents.

2. Mary had a cake which she cut into 10 pieces; 7 were eaten. How many were left?

3. My knife has 6 blades; 2 of them are open. How many are closed?

See P. Ed., p. 9.

4. James bought a paper for 5 cents ; he gave the newsboy 10 cents. How much change should James receive?

5. Samuel put 9 peaches on the table and his sister took 5 of them ; how many were left?

6. A man owed $7 ; he paid $3 ; how many $'s did he then owe?

7. Willis took 9 cents to buy candy with ; he lost 4 cents. How many had he to buy candy with?

8. Henry bought a pencil for 4 cents and sold it for 7 cents ; how many cents did he gain?

8. Matthew bought one pencil for 4 cents and another for 5 cents ; what did both cost him?

10. Susan bought 3 spools of white thread and 6 spools of blue thread ; how many spools did she buy?

11. Jane tried to solve 6 examples ; she had 4 of them correct. How many were wrong?

12. Walter had 8 pencils ; he broke 3 of them. How many whole ones had he?

13. Fanny had 6 needles ; she found 4 more. How many had she then?

14. May is 7 years old and her brother

Frank is 4 years old ; how much older is May than Frank?

15. George had a stick 9 inches long ; he cut off 3 inches of it. How long was the stick then?

16. A farmer having 8 turkeys, sold 4 of them ; how many had he left?

17. John paid 3 cents for candy and 5 cents for marbles : how many cents did he spend?

18. A little boy had 3 fingers cut off in a machine ; how many had he left?

19. Silas had 3 marbles in one pocket and 5 in the other ; how many had he in both?

20. How many wheels have a sulky and a wagon together?

21. There are 3 girls on the front seat of a carriage, and 5 girls on the back seat ; how many girls in the carriage?

22. Jesse had 8 sticks of wood to bring in ; he has brought in 3 ; how many more has he to bring in?

23. An orange cost 6 cents, and a peach cost 3 cents ; how much more did the orange cost than the peach?

24. If a pear cost 4 cents, and a lemon cost 5 cents, what will a pear and a lemon cost?

25. There were 6 eggs in a nest and 4 of
See P. Ed., p. 11.

them were broken ; how many whole ones were there?

26. Ellen's father gave her 9 cents ; she bought a doll with 5 cents. How many cents had she left?

27. There are 4 boys riding in a sleigh and 2 riding behind on the runners ; how many boys with the sleigh?

28. Ella has 5 roses on her bush, and 5 in her hand ; how many has she?

29. There were 9 chickens in a coop and a rat ate 3 of them ; how many were left?

30. There are in the class 4 girls and 3 boys ; how many pupils in the class?

31. A little boy bought 10 sticks of candy ; he ate 3 of them and gave away the rest. How many did he give away?

These 31 examples should be reviewed and others given, until the pupils know at once in such simple problems whether they are to find the sum or the difference. Review the series also.

Count by 4's from 1 to 17.

Copy the following examples one at a time on the blackboard ; require a pupil to point one off into periods and read it. Erase it, then

FIRST STEPS AMONG FIGURES. 69

write another and require another pupil to point it off and read and so with the others :

1. 90016. 2. 45378. 3 461340.
4. 908714. 5. 876341. 6. 608790.
7. 379000. 8. 75608. 9. 40713.
10. 100740. 11. 98716. 12. 900000.
13. 800601. 14. 200003. 15. 761300.
16. 500000. 17. 700300. 18. 60050.
19. 700060. 20. 600000. 21. 200361.
22. 500700. 23. 40010. 24 900007.

Teach Arabic notation to 999,000.

To be read by the teacher for pupil to write upon slates or blackboard.

Write in Arabic the following :

1. 1,040. 2. 3,506. 3. 10,016.
4. 8,400. 5. 9,350. 6. 7,518.
7. 3,761. 8. 10,010. 9. 40,070.
10. 73,801. 11. 36,000. 12. 90,090.
13. 100,100. 14. 702,940. 15. 900,070.
16. 816,902. 17. 49,049. 18. 860,705.
19. 461,017. 20. 791,486.21. 21. 10,020.

Write these, or similar numbers on the board and require the pupils to read them.

Write in Roman the following :

1. 79. 2. 96. 3. 110.
4. 47. 5. 19. 6. 134.

See P. Ed., p. 26.

Write in Arabic:
7. CLX.   8. XCII.   9. CXC.
10. CXLIV.   11. CLXXXVI.   12. CIX.

Write in Roman :
13. 192.   14. 136.   15. 168.   16. 154.

A short method of adding equal numbers is called Multiplication;

or,

Taking a number a certain number of times is called Multiplication.

The number obtained by multiplication is called the Product.

1. John bought 5 pencils at 4 cents each; what did they cost?

Solution : They cost 5 times 4 cents, or 20 cents ; or if *one* pencil cost 4 cents, *five* pencils will cost 5 times 4 cents, or 20 cents.

2. If 1 orange cost 5 cents, what will 4 oranges cost?

3. What cost 4 marbles at 3 cents each?

4. How many quarts in 2 gallons?

Solution : In one gallon there are 4 quarts, in 2 gallons there are 2 times 4 quarts, or 8 quarts.

5. How many pints in 3 quarts?

6. How many quarts in 5 gallons?

7. How many wheels have 3 wagons?

8. A lady gave 3 little girls 5 bunches of grapes each; how many bunches did she give them all?

9. How many feet have 4 hens.

10. How many feet have 3 cows?

11. What cost 5 books at 4 shillings each?

12. What cost 3 lead pencils at 5 cents each?

13. If a lead pencil cost 6 cents and a marble cost 3 cents, what will both cost?

14. What cost a doll worth 4 cents, and a spool of thread worth 6 cents?

15. What cost 3 pencils at 4 cents each?

16. The boys are riding down hill on sleds; there are 4 sleds and 2 boys on each sled; how many are riding down hill?

17. A boy bought a sled for 8 shillings and sold it for 5 shillings; how many shillings did he lose?

18. A boy had 8 cents in his pocket, but he lost 4 of them through a hole in his pocket; how many had he left?

19. A boy paid 4 cents for candy and 5 cents for nuts; how much money did he spend?

Review these examples carefully.

Count by 4's from 1 to 29.

For rapid solving.

See P. Ed., p. 19.

$12+4+3+3+3+4+4+4+4+3+3+3$
$+2+2=?$ Ans. 41.

$3+4+4+3+3+3+3+2+2+2+3+3$
$+2+3+4=?$ Ans. 44.

$3+3+2+2+2+3+3+3+2+3+2+2$
$+3+3+3=?$ Ans. 39

$3+4+4+4+3+3+2+2+4+3+3+2$
$+2+2=?$ Ans. 41.

$2+3+3+3+2+2+2+2+3+3+2+3$
$+3+4=?$ Ans. 37.

$3+2+2+2+3+3+2+2+1+3+2+3$
$+3+1=?$ Ans. 32.

$4+3+4+4-3+4-3-2+4+4-3+2$
$+3=?$ Ans. 21.

$2+3+4+4+4-3+2-3-2-3+4+2$
$+3=?$ Ans. 17.

$2+4+3+4+3-4-3-4+3+3+4+4$
$+3=?$ Ans. 22.

$5+3-2-4+3+3+2+4+3+4+3-2$
$-3=?$ Ans. 19.

$3+4+2+4+2+2-4-4+3-2-4+3$
$+4=?$ Ans. 13.

$3+2+4-3-2+3+3+3+3+4-3-2$
$-4+3+2=?$ Ans. 16.

$4+3+2+3+3+4+3-4-4+1-3+4$
$-3-4-4=?$ Ans. 5.

20. What cost 5 marbles at 4 cents each?

21. What cost a marble at 3 cents and a pencil at 4 cents?

22. James bought 3 books at 4 shillings each; what did they cost him?

23. David had 8 apples when he started for school, but he ate 3 on the way; how many had he when he got to school?

24. Sarah ate 3 crackers at breakfast and 5 at dinner; how many did she eat at both meals?

25. How many horses in 3 four-horse teams?

Finding how many times one number is contained in another is called Division.

The number which shows how many times it is contained is called a Quotient.

1. How many pears at 2 cents each can be bought for 8 cents?

Solution: If 1 pear cost 2 cents, for 8 cents you can buy as many pears as 2 is contained times in 8, or 4; or, as many as there are 2's in 8, or 4.

2. John has 15 cents; how many marbles at 3 cents each can he buy?

3. Willis spent 20 cents for oranges at 5 cents each; how many oranges did he buy?

See P. Ed., p. 25.

4. If one pig cost $4, how many pigs may be bought for $12?

5. How many lead pencils at 4 cents each can be bought for 20 cents?

6. If 1 doll cost 3 shillings, how many such dolls can be bought for 12 shillings?

7. 10 shillings will buy how many knives at 2 shillings each?

8. How many balls at 3 shillings each may be bought for 6 shillings?

9. When pears are 2 cents each, how many can you buy for 8 cents?

Count by 4's from 1 to 61.
" " 4's " 2 " 62.
" " 4's " 4 " 60.
" " 4's " 3 " 63.

10. What cost 4 pineapples at 2 shillings each?

11. How many pencils at 4 cents each can be bought for 16 cents?

12. A boy walked 5 miles 1 day, and 3 miles the next day; how far did he walk in the 2 days?

13. How many pairs of mittens at 3 shillings a pair can you buy for 9 shillings?

14. 20 shillings will buy how many purses at 5 shillings each?

15. Henry earned 9 cents on Monday and 5 cents on Tuesday; how many cents more did he earn on Monday than on Tuesday?

16. Louisa had 5 cents and she found 3 cents more; how many cents had she then?

17. If one sled cost 5 shillings how many sleds can you buy for 20 shillings?

18. What cost 4 vests at $8 each?

19. How many neckties at 3 shillings each can you buy for 12 shillings?

20. 5 little boys each have a pair of copper-toed boots; how many boots have they?

Review these examples.

For addition and multiplication. (6 and rev.)

a	b	c	d	e
3 6 2 4	1 3 6 2	4 1 5 4	6 2 4 1	5 5 3 6
2 3 4 5	6 3 4 5	6 2 3 4	5 6 2 3	2 4 5 6

f	g	h
2 4 1 5	3 6 2	3 1 5
2 3 4 5	6 2 3	4 5 6

For subtraction.

a	b	c	d
9 7 6 10	7 11 7 8	3 10 8 6	4 7 9 8
5 6 3 4	5 5 4 3	2 6 6 1	3 2 4 5

e	f	g	h
12 4 7 11	6 8 5 8	9 10 5	5 9 6
6 2 3 6	5 4 3 2	6 5 4	2 3 4

See P. Ed., p. 28.

For division.

a				b				c			
30	12	8	3	10	20	15	36	12	15	2	24
5	6	2	3	2	4	5	6	4	3	2	6

d				e				f			
4	12	4	25	18	12	6	16	5	30	6	18
2	3	4	5	6	2	3	4	5	6	2	3

g			h		
8	20	6	9	24	10
4	5	6	3	4	5

(1)	(2)	(3)	(4)	(5)	(6)
32	33	2	23	33	31
13	21	32	33	31	23
3	2	23	31	23	33
21	30	33	12	13	23
13	13	21	22	22	32
33	32	23	31	32	13
115	131	134	152	154	155

(7)	(8)	(9)	(10)	(11)	(12)
30	23	32	3	31	23
23	33	30	23	33	32
33	30	23	31	23	13
12	22	3	23	30	33
31	31	31	33	22	21
23	33	23	32	31	32
33	22	33	21	33	23
185	194	175	166	203	177

See P. Ed., p. 16.

13. Add 32, 21, 3, 33, 23, 31, 23, 33.
14. Add 33, 23, 30, 21, 2, 33, 23, 22.
15. Add 2, 33, 22, 31, 23, 32, 12, 23.
16. Add 31, 30, 23, 33, 21, 32, 22, 32.
17. Add 23, 13, 32, 23, 3, 13, 32, 33.

Count by 5's from 5 to 60.

For rapid solving.

$3 \div 4 + 4 + 3 + 3 + 2 + 4 + 3 + 3 + 2 + 4 + 4 + 3$? Ans. 42.

$2 + 3 + 4 + 3 + 3 + 5 + 5 + 3 + 3 + 4 + 4 + 4 + 3 + 3 + 2$? Ans. 51.

$3 + 4 + 4 + 3 + 3 + 4 + 4 + 5 + 5 + 5 + 3 + 2 + 4 + 4 + 4$? Ans. 57.

$3 + 4 + 3 + 3 + 3 + 2 + 2 + 3 + 4 + 2 + 4 + 3 + 2 + 3 + 5$? Ans. 46.

$5 + 3 + 4 + 2 + 3 + 4 + 3 + 3 + 2 + 4 + 3 + 4 + 3 + 3 + 4$? Ans. 50.

$4 + 3 + 4 + 3 - 2 - 2 + 5 - 3 + 2 + 4$? Ans. 18.

$2 + 4 \div 3 \times 5 - 4 \times 4 - 3 - 3 \div 3 + 6 \div 4 \times 6 + 4 + 3 - 2$? Ans. 23.

$3 + 6 + 4 + 4 + 4 + 3 + 4 - 6 - 4 - 2 - 5 + 3 + 4 - 4$? Ans. 14.

$3 + 5 \div 2 \times 5 - 3 + 4 - 2 - 4 \div 5 \times 6 + 4 + 3 \div 5 + 3 + 3$? Ans. 11.

2+5+4+1÷4+1×5—4+3 +4+1÷4
+5+2+4+3? Ans. 20.
3+6+4—1 — 3×6—3—4—3 +4÷6×5
+4+3+4+3? Ans. 29.
5+6—3+4+3+3 ÷ 3 + 4+3+4+4—2
—3+4+3+4? Ans. 27.

Do not use all of these at once, but use them occasionally and in connection with a lesson in examples of another kind, or to wake up the whole school sometimes when they are listless.

1. What cost a pencil at 6 cents and a marble at 4 cents?

2. Mary had 5 peaches and her brother gave her 6 more; how many had she then?

3. What cost 5 books at 6 shillings each?

4. 3 boats are on the lake; each has a pair of oars, how many oars have the 3 boats?

5. There are 9 boys in a class, and 6 of them recite well; how many do not recite well?

6. How many baskets, at 4 cents each, can be bought for 24 cents?

7. Charles spent 18 cents for candy at three cents an ounce; how many ounces did he buy?

8. Jane had 10 needles, she lost 7 of them; how many had she then?

9. A lazy boy brought his mother 3 sticks of wood at one time and 4 at another; how many sticks did he bring her?

10. Six boys can sit on this seat, how many boys can sit on four such seats?

11. In a school room there are 6 keys hanging on a nail, 2 keys for each door; how many doors are there?

12. There are 18 words in the spelling lesson, 3 words for each pupil; how many pupils in the class?

13 There are 5 piles of books and 6 books in each pile; how many books in the 5 piles?

14. One stormy day George cleared the path of snow 4 times in the forenoon and 5 times in the afternoon; how many times did he clear the path?

15. Amelia had 11 cents and spent 5 of them; how many had she left?

16. Arthur had 7 buttons on his jacket; how many had he after losing 2 of them?

17. How many quarts in 6 gallons?

18. Mr. Smith has a quart of maple syrup; how many times can he fill a pint cup with it?

19. How many skates at 6 shillings each can you buy for 24 shillings?

20. What cost a knife at 5 shillings and a saw at 6 shillings?

21. If 2 oranges cost 12 cents, what will 1 orange cost?

Solution: If 2 oranges cost 12 cents, 1 orange will cost ½ of 12 cents, or 6 cents; or 1 orange will cost ½ of 12 cents, or 6 cents.

Before giving examples like the above teach the pupils carefully that if 2 things of equal value cost a certain sum, 1 of them will cost ½ of that sum; if 3 cost that sum, 1 of them will cost 1-3 of it; if 5 cost any sum, 1 of them will cost 1-5 of it; if 9 of them cost any sum, 1 of them will cost 1-9 of it, &c.

Question on this subject until it is thoroughly mastered.

Show the pupils that to get ½ of 12 apples (or marbles or pencils) they may be placed in 2 equal piles, and they will find that ½ of 12 apples is 6 apples.

Show them that to get ⅓ of 12 pencils, they may be placed in 3 piles, and that ⅓ of 12 is 4. Show in the same way that ¼ of 12 is 3. When this is thoroughly understood, show them that ½ of 12 may be obtained by dividing 12 by 2 — the result in each case being 6; show that ⅓

FIRST STEPS AMONG FIGURES.

of ½ may be obtained by dividing 12 by 3; teach ¼ of 12 in like manner. Illustrate also by ½ of 6 and ⅓ of 6 and ¼ of 8.

Then teach in general terms that ½ of any number may be obtained by dividing the number by 2; ⅓, by dividing by 3; ⅐, by 7, &c.

22. What cost 1 pear if 4 pears cost 8 cents?

23. If 3 knives cost 15 shillings what will 1 knife cost?

24. If 2 pencils cost 16 cents, what will 1 pencil cost?

25. If 4 stools have 12 legs, how many legs will 1 stool have?

26. If 6 boys earn 18 cents, how many does 1 boy earn?

27. If 5 cords of wood cost $25, what will 1 cord cost?

28. How many pounds in 1 box of honey if 4 boxes contain 24 pounds?

29. At 5 cents each, how many oranges can be bought for 30 cents?

30. If 3 lemons cost 18 cents, what costs 1 lemon?

31. How many pounds of butter will last a family 1 week if they use 12 pounds in 4 weeks?

32. How many days will 18 apples last a boy who eats 3 apples each day?

33. A blacksmith shod 5 horses each day; how many did he shoe in 6 days?

34. 6 boys are skating on the ice, and 4 boys are sliding on the ice without skates; how many boys on the ice?

35. Nine boys were riding down hill on sleds; 3 of them went home. How many continued to ride down hill?

36. If 15 yards of cloth will make 5 pairs of trowsers, how many yards will it take to make 1 pair of trowsers?

37. George has 4 books, and Mary has 5 books; how many have both?

38. 12 cents will buy how many marbles at 3 cents each?

39. If 5 marbles cost 10 cents, what will 1 marble cost?

40. What cost 1 apple if 5 apples cost 10 cents?

41. How many dolls at 4 shillings can you buy for 20 shillings?

Review the last 21 examples until the pupils solve them readily and can distinguish when they divide and when get one-half or one-third, &c.

Count by 5's from 1 to 61.

Count by 4's from 1 to 61.
Count by 4's from 2 to 62.
Count by 3's from 3 to 60.
Count by 3's from 1 to 61.
Count by 3's from 2 to 62.
To be read by the teacher.
Write in Arabic:

1. 7,050.   2. 10,003.   3. 40,300.
4. 5,209.   5. 10,010.   6. 300,040.
7. 9,610.   8. 4,316.   9. 215,000.
10. 80,090. 11. 600,000. 12. 809,740.
13. 100,010. 14. 916,008. 15. 835,941.
16. 70,000. 17. 90,005. 18. 5,016.
19. 213,033. 20. 30,000.

Write these or similar numbers on the blackboard and require pupils to read them.

For rapid solving.

$6+4+3+2+4+4-3 \div 4 \times 6-3-3+4+4$? Ans. 32.

$3+5+4+3+3+4+3 \div 5 \times 4-4+3+3-4+3+5+5$? Ans. 31.

$4+3+4+4+4+2+3+4+4+2+5+2+3+3+4$? Ans. 53.

$3+2+4+3+4+3+2+3+3+4+3+2+3+3+2$? Ans. 44.

$2+3+3+2+4+4+3+2+3+3+4+4+2+3+3+4+3$? Ans. 52.

FIRST STEPS AMONG FIGURES.

$3+2+4+3+5+3+2+4+4+1+2+4+1+3+1+2$? Ans. 44.

$4+3+3+4+2+3+2+3+3+3+2+4+1+2$? Ans. 39.

$3+2+4+4+3+3+2+2+3+4+2+3+3+2+1$? Ans. 41.

$17+4+3 \div 6+6+5-3 \div 3 \times 6+3-4-3 \div 5 \times 3$? Ans. 12.

$9+4+3+2 \div 3 \times 5-4-3-2+4-1 \div 6 \times 3+3+2$? Ans. 17.

$16+4+3+3+3+2-1 \div 5+3+3 \div 3 \times 5-3+4-1 \div 4$? Ans. 5.

In giving the following examples as well as those "for rapid solving" the teacher should be very careful that pupils do not acquire a pernicious habit of counting instead of adding at sight or as soon as heard.

Slate examples.

(1.)	(2)	(3)	(4)	(5)	(6)	(7)	(8)	(9)
2	1	2	2	1	2	1	2	1
1	2	1	2	2	1	2	1	2
2	2	1	1	2	2	2	2	2
2	1	2	2	1	2	2	2	2
1	2	1	2	2	2	2	2	2
2	2	2	2	1	2	1	2	1
1	2	2	2	2	2	2	2	2
—	1	1	2	2	2	2	1	2
11	—	—	1	1	1	1	2	1
	13	12	—	—	—	—	—	—
			16	14	16	15	16	15

FIRST STEPS AMONG FIGURES.　85

(10)	(11)	(12)	(13)	(14)	(15)	(16)	(17)	(18)	(19)
2	1	2	2	2	2	1	1	2	1
1	2	1	1	2	1	2	2	2	2
2	2	2	2	2	1	2	2	1	2
2	2	2	2	2	1	2	1	2	2
2	1	2	2	1	1	1	2	2	1
1	2	1	2	2	2	2	1	1	1
2	2	2	1	2	2	1	1	1	1
1	2	2	2	1	1	2	2	1	1
2	1	1	1	2	2	2	2	2	2
1	2	2	2	2	2	1	1	2	2
2	2	2	2	2	1	1	1	1	2
2	2	2	1	2	2	1	2	1	1
1	1	1	1	2	2	1	1	—	—
—	—	—	—	—	—	—	—	18	18
21	22	22	21	24	20	19	19		

(20)	(21)	(22)	(23)	(24)	(25)
2	1	1	2	2	1
1	2	2	1	1	2
2	2	2	2	2	2
2	2	1	2	2	2
2	1	2	2	1	2
1	1	1	1	1	2
1	2	2	1	2	1
2	1	2	1	2	1
2	1	2	2	2	1
2	2	2	2	1	2
2	2	1	1	1	2
2	1	2	1	1	2
1	2	—	1	1	1
—	—	20	—	—	—
22	20		19	19	22

(26)	(27)	(28)	(29)	(30)	(31)	(32)
2	1	2	1	2	2	2
1	2	1	2	1	1	1
2	1	2	1	2	2	2
2	1	2	2	2	2	1
2	2	1	1	2	2	2
2	1	2	2	1	1	2
1	1	1	1	2	2	2
2	1	2	1	1	1	2
1	2	1	1	2	1	2
2	2	2	2	2	2	1
2	1	2	1	1	2	2
2	1	2	2	2	1	1
2	2	1	1	2	2	2
1	2	2	2	2	2	2
—	1	1	1	—	2	2
24	—	—	—	24	—	—
	21	24	21		25	26

(33)	(34)	(35)	(36)	(37)	(38)	(39)	(40)
1	2	1	2	1	2	1	2
2	1	2	1	2	1	2	1
2	2	1	2	1	2	1	1
2	1	1	1	2	2	1	1
1	1	2	1	2	2	1	2
2	1	1	2	2	1	2	2
1	2	2	2	1	2	2	1
1	2	2	1	2	2	1	2
2	2	2	2	2	2	2	2
2	2	1	2	2	2	2	2
2	1	1	1	1	1	2	1
2	1	1	1	2	2	1	2
1	2	2	1	2	2	2	1
2	1	1	2	1	1	2	2
1	2	2	2	2	2	2	1
2	1	1	—	2	1	1	—
—	—	—	23	—	—	—	23
26	24	23		27	27	25	

If the teacher is careful that the pupils do not keep any of the solutions of the foregoing examples, they may be given 2 or 3 times over, first solving them all, then solving them all again, and so on.

FIRST STEPS AMONG FIGURES.

(41)	(42)	(43)	(44)	(45)	(46)	(47)	(48)	(49)
2	23	30	13	32	23	32	33	13
3	32	23	21	13	33	20	21	21
1	12	31	33	32	30	33	13	33
2	21	12	10	21	12	23	23	23
1	30	23	32	33	33	31	32	32
3	12	32	13	20	22	23	33	13
3	31	33	22	32	31	32	22	22
2	23	21	31	23	13	11	12	31
3	33	20	23	33	33	33	31	20
20	217	225	198	239	230	238	220	208

(50)	(51)	(52)	(53)	(54)	(55)	(56)	(57)
32	21	31	23	30	23	33	23
21	33	23	31	23	31	21	31
33	20	32	33	10	23	32	23
23	13	20	30	22	30	13	32
12	32	13	22	31	22	30	33
31	33	31	13	33	13	22	12
33	21	23	31	23	32	33	31
23	13	32	23	12	21	21	23
30	32	13	30	33	33	13	32
23	31	23	23	22	23	32	33
13	33	33	33	32	32	33	23
274	282	274	292	271	283	283	296

See P. Ed., p. 33.

FIRST STEPS AMONG FIGURES. 89

(58)	(59)	(60)	(61)	(62)	(63)	(64)	(65)	(66)
21	33	23	31	33	13	32	21	33
33	23	31	23	21	22	13	33	22
20	31	13	33	13	31	31	32	31
13	20	32	20	30	33	23	23	23
32	32	20	13	23	21	33	30	31
23	13	33	32	32	13	21	22	12
32	33	21	23	11	32	13	33	23
21	21	13	12	33	33	10	12	32
13	32	32	22	23	20	32	32	31
32	12	22	33	32	13	31	23	23
23	33	30	30	33	32	23	31	13
31	21	13	21	21	23	32	13	33
12	23	32	33	33	31	31	23	32
306	327	315	326	338	317	325	328	339

Teach pupils to prove every example in addition by adding both upward and downward, and in this way they will get more practice—just what is needed.

If the pupils have their books—P. Ed.—the following examples are intended to be given at recitation for immediate solution, while those in the P. Ed. may be solved by the pupils at their seats and brought to recitation.

FIRST STEPS AMONG FIGURES.

If the pupils have not got their books these examples may be written on the board or read to pupils to solve at their seats, a few daily:

(1)	(2)	(3)	(4)	(5)	(6)
231	321	333	331	132	133
322	233	223	223	223	332
122	213	312	113	331	123
333	332	233	322	233	322
212	322	213	232	213	331
323	123	322	131	332	213
332	332	132	303	213	321
221	233	323	233	233	33
2,096	2,109	2,091	1,888	1,910	1,808

(7)	(8)	(9)	(10)	(11)	(12)
33	313	213	123	221	232
203	232	2	331	323	333
332	231	321	322	332	123
123	323	233	233	121	232
333	333	332	123	203	313
212	232	23	312	323	233
332	321	231	233	232	321
323	233	133	331	123	333
223	133	213	223	333	231
331	312	321	312	312	312
2,445	2,663	2,022	2,543	2,523	2,663

Review these if need be; in any case be sure the pupils can add such examples as the above *readily* and accurately.

See P. Ed., p. 36.

Examples in subtraction.

13. 69758	14. 97856	15. 69587	16. 68059
6035	74230	20152	6040
63,723	23,626	49,435	62,019

17. 75860	18. 79685	19. 58796	20. 96807
2330	4242	25062	4200
73,530	75,443	33,734	92,607

Multiplication.

21. 32032	22. 23103	23. 24130	24. 31402
2	3	2	2
64,064	69,309	48,260	62,804

25. 23103	26. 14023	27. 32023	28. 20312
2	2	3	3
46,206	28,046	96,069	60,936

29. 40312	30. 31203
2	3
80,624	93,609

Division.

31. 2)48206   32. 3)90396   33. 3)69306
    ———   ———   ———
     24,103   30,132   23,102

34. 2)28460   35. 3)39069   36. 2)60482
    ———   ———   ———
     14,230   13,023   30,241

37. 2)84602   38. 3)30960   39. 3)93600
    ———   ———   ———
     42,301   10,320   31,200

(1)	(2)	(3)	(4)	(5)	(6)
324	434	23	43	342	431
431	431	431	324	434	343
243	444	344	442	44	242
432	213	244	243	321	434
444	332	432	134	443	244
342	441	421	424	2	421
232	424	444	233	314	314
2,448	2,719	2,339	1,843	1,900	2,429

See P. Ed., p. 38.

FIRST STEPS AMONG FIGURES. 93

(7)	(8)	(9)	(10)	(11)	(12)	(13)
424	31	422	431	21	34	414
434	42	344	323	44	44	443
342	44	142	444	14	21	302
241	4	433	444	23	30	244
424	31	424	123	42	44	431
332	43	344	334	44	23	444
2,197	195	2,109	2,099	188	196	2,278

(14)	(15)	(16)	(17)	(18)	(19)	(20)
214	44	44	342	244	21	23
424	213	23	444	434	34	41
444	321	32	32	423	43	34
123	444	41	421	342	21	44
434	3	44	344	231	44	31
341	432	31	432	444	32	23
344	424	22	444	234	43	42
2,324	1,881	237	2,459	2,352	238	238

(21)	(22)	(23)	(24)	(25)	(26)	(27)
444	21	423	432	34	442	24
344	43	444	444	42	434	43
213	24	144	324	33	324	21
422	44	231	131	23	413	4
343	32	432	423	44	32	32
214	24	344	442	3	424	20
444	33	432	343	22	321	44
321	44	123	414	31	444	23
2,745	265	2,573	2,953	232	2,834	211

FIRST STEPS AMONG FIGURES.

(28)	(29)	(30)	(31)	(32)	(33)	(34)
243	21	44	423	341	44	34
444	432	21	341	444	21	42
231	344	34	444	34	4	44
422	211	43	332	432	44	13
444	321	43	213	121	32	22
233	432	31	441	343	12	41
423	343	24	324	444	33	44
344	444	32	432	434	14	34
212	323	43	344	232	43	31
440	342	44	443	341	44	42
3,436	3,213	359	3,737	3,166	291	347

(35)	(36)	(37)	(38)	(39)	(40)	(41)
343	423	14	314	232	21	32
444	442	3	432	443	42	443
321	341	41	341	444	43	344
432	234	34	134	324	14	214
344	434	42	244	431	44	423
344	341	44	441	343	21	4
234	423	34	342	234	13	234
421	412	31	444	444	44	342
432	344	23	233	323	42	444
444	431	42	324	242	33	431
3,759	3,825	308	3,249	3,460	317	2,911

See P. Ed., p. 42.

FIRST STEPS AMONG FIGURES.

(42)	(43)	(44)	(45)	(46)	(47)	(48)
432	34	341	24	234	32	231
343	41	423	43	421	44	443
231	4	444	32	444	43	422
424	23	34	44	324	24	334
344	42	123	4	413	31	244
243	13	442	33	442	44	413
412	24	344	24	134	43	423
424	44	431	41	224	24	341
243	33	233	43	442	31	444
432	21	342	34	431	44	413
234	43	444	42	344	13	244
3,762	322	3,601	364	3,853	373	3,952

For addition and multiplication. (7 and rev.)

a	b	c	d	e
4 7 3 5	2 4 7 3	2 6 4 7	3 5 2 6	4 7 3 5
3 4 5 6	7 4 5 6	3 4 5 6	7 3 4 5	6 7 3 4

f	g	h
2 6 4 7	3 5 2	6 6 5
5 6 7 3	4 5 6	7 3 7

For subtraction.

a	b	c	d
11 9 8 12	10 5 9 14	10 6 9 7	12 11 10 7
6 7 4 5	4 3 6 7	6 3 4 5	6 7 3 4

e	f	g	
10 7 11 8	12 9 13 8	9 8 10 6 11	13
5 3 4 5	7 3 7 6	5 3 7 4 5	6

For division.

```
         a           |        b          |         c
   12 25 12 21       |  28 36 49 24      |   9 20 10 42
    6  5  4  3       |   7  6  7  6      |   3  4  5  7

         d           |        e          |         f
   18 35 15  8       |  30 21 42  6      |  24 20 18 35
    3  7  3  4       |   5  7  6  3      |  .4  5  6  5

                         g        |      h
                     16 14 12     |  28 15 30
                      4  7  3     |   4  5  6
```

(1)	(2)	(3)	(4)	(5)
342	24	243	404	23
421	32	434	34	34
34	44	241	423	44
430	13	343	342	3
244	40	234	424	43
423	34	413	334	21
342	21	324	41	34
244	44	34	423	44
321	32	443	342	32
143	42	3	333	43
432	33	234	423	4
341	4	444	342	44
132	13	321	244	32
324	44	243	344	43
431	32	342	321	32
4,604	452	4,296	4,774	476

See P. Ed., p. 40.

6. Add 342, 234, 344, 421, 342, 24, 431, 231, 4, 423, 344, 434, 324.

7. Add 24, 43, 24, 32, 44, 34, 42, 4, 34, 21, 43, 14, 34, 23, 43.

8. Add 43, 233, 424, 341, 3, 434, 4, 342, 434, 342, 243, 414.

9. Add 2, 34, 44, 41, 32, 43, 34, 3, 44, 34, 23, 42, 34, 41.

How many quarts are there in three gallons?

Solution: In *one* gallon there are 4 quarts, in *three* gallons there are three times 4 quarts or 12 quarts.

The following examples may be given during recitation:

1. How many feet have 5 horses?

2. Arthur was paid 5 cents for doing an errand and his sister gave him 4 cents; how many had he then?

3. If a carpenter can drive 3 nails in a minute, how many minutes will it take him to drive 18 nails?

4. Charles had 10 snow balls in a pile; he threw 4 of them at his playmates. How many remained in the pile?

5. Mr. Smith paid 4 dollars for the cloth for a pair of pants and 2 dollars for making them ; what did the pants cost him?

6. How many vests at $4 each can be bought for $24?

7. At 4 shillings a pair, what cost 5 pairs of scissors?

8. If a hat cost $3 and a pair of boots $10, how much more do the boots cost than the hat?

9. Andrew has a pair of ponies, how many feet have they?

10. How many more feet than eyes has a four-horse team?

11. How many less heads than feet has a three-horse team?

12. A boy spent 21 cents for marbles at 3 cents each; how many marbles did he get?

13. Barton has 17 cents ; how many pencils at 2 cents each can he buy and keep 3 cents?

$3+5+4+5+5+3+4+5+3+4+2+3+2+4=$? Ans. 52.

$5+5+3+3+4+4+5+3+2+4+1+5+4+5+3+4=$? Ans. 60.

Examples like the above, having only addition, may be given both forward and backward, thus they will make 4 examples instead
See P. Ed., p. 45.

FIRST STEPS AMONG FIGURES. 99

of 2. Still more may be made by commencing with another number, as 12, and adding 12 to the answer. Thus:

$12 + (5+5+3+3+4+4+5+3+2+4+1+5+4+5+3+4) = 60+12 = 72$.

$8+4+5+4 \div 3 \times 5+4+4+5-3-4-5-4-5-3-3$? Ans. 21.

$15+4+4-5 \div 3+5+5+4 \div 4+7+3+4+5+5$? Ans. 29.

$4+5+3+4+2+5+4+5+3+2+5+5+4+4+3$? Ans. 58.

$3+4+5+3+1+5+4+5+2+3+5+4+5+4+4+5$? Ans. 62.

$8+5+4+5+2+4 \div 4 \times 3 - 5 \div 4 \times 8 - 4 - 5 - 5 - 4$? Ans. 14.

$16+4+5+3+4-4 \div 7 \times 6+5-3-5-4+2+5 \div 6+5$? Ans. 9.

$18+5+4+4+5-3-4-5 \div 4+5+5 \div 4 \times 5+4+4 \div 4$? Ans. 7.

$3+2+4+5+5+3+4+4+2+5+3+4+5+3+3+2+4$? Ans. 61.

$4+5+4+3 \div 4 \times 7+5-2-3-4-5-3-5-4 \times 3+3 \div 6$? Ans. 4.

$4+2+5+3+3+5+4+4+5+5+3+2+4+1+5+4 | 5+3$? Ans. 67.

$3+4+4+5+2+5+3+3+4+5+4+4+5+5+5+3+4$? Ans. 68.

FIRST STEPS AMONG FIGURES.

$3+4+5+3+4+5+5+4+4+3+5+2+5+3+5+4+5$? Ans. 69.

$5+4+5+4+5+1 \div 6 +3+5+4+5-3-5-1 \div 4 \times 7-5-4$? Ans. 12.

$7+4+5+5+4+4+3-5-4-5-3-4-4 \times 3$? Ans. 21.

$5+3+5+4+4+3+3+5+5-3-3 \div 4-5-5-4-4-3$? Ans. 6.

(1)	(2)	(3)	(4)	(5)	(6)	(7)
35	5	54	50	34	45	32
54	43	35	35	54	53	45
5	54	43	4	45	24	53
43	35	5	53	33	45	34
24	42	52	45	54	51	15
55	51	44	24	44	35	52
34	35	25	55	32	54	41
52	54	53	31	55	5	35
45	45	21	44	34	32	54
347	364	332	341	385	344	361

Examples in subtraction:

1. 67,548
    43,235
    ──────
    24,313

2. 69,584
    34,331
    ──────
    34,253

3. 75,897
    54,353
    ──────
    21,544

4. 97,867
    52,343
    ──────
    45,524

5. 64,786
    2,432
    ──────
    62,354

6. 79,684
    7,053
    ──────
    72,631

See P. Ed., p. 47.

FIRST STEPS AMONG FIGURES. 101

The teacher may use his own judgment as to teaching subtraction when some figures of the subtrahend are greater than the corresponding figures of the minuend.

A purely mechanical method is here given with the idea that the *method* of doing many things may properly precede the *reason* for the method.

If the following method be used, after one or two years or in a larger book the reason of the method should be fully explained to the pupil, and he should then be required to give the reasoning himself.

32,413 — 5,667. Solve by separating the figures of the minuend, as in the line below, and then when any figure of the subtrahend is smaller that the figure above it, write 1 before it thus:

$$\begin{array}{ccccc} 3 & {}^{1}2 & {}^{1}4 & {}^{1}1 & {}^{1}3 \\ & 5 & 6 & 6 & 7 \\ \hline 2 & 6 & 7 & 4 & 6 \end{array}$$

and then subtract, being careful when 1 is prefixed to the upper figure to add 1 to the next left hand figure of the subtrahend. The following examples are so arranged that no figure of the subtrahend is greater than 7, the tables having been learned only so far. After solving the above example the pupil should say 5,667 from 32,413 leaves 26,746.

7. 910201	8. 423423	9. 621423
56454	45466	53667
853747	377957	567756

As soon as the pupil understands the mechanical work, he should not be allowed to write the 1's in the minuend, but imagine them to be there.

10. 831,242	11. 513,423	12. 430,213
63,567	46,647	262,637
767,675	466,776	167,576

13. 731,420	14. 342,031	15. 532,514
54,654	25,266	65,251
676,766	316,765	467,263

16. 731,420	17. 624,091	18. 831,042
61,265	62,035	63,415
670,155	562,056	767,627

See P. Ed., p. 48.

FIRST STEPS AMONG FIGURES. 103.

(1)	(2)	(3)	(4)	(5)	(6)
24	532	523	24	355	544
55	455	454	55	432	352
32	44	342	44	534	435
44	503	535	32	255	534
53	434	453	53	543	253
25	355	544	45	424	545
43	3	325	54	354	324
5	545	454	34	535	432
54	334	543	43	444	543
335	3,205	4,173	384	3,876	3,962

More examples may be made from these by reading them from the center each way, thus giving new combinations, or by giving two additional numbers, one above the upper number and one below the lower one; in this way the combinations will be different whether the pupil add upward or downward.

Of course the teacher must add the sum of these two numbers to the answer in the book to get the answers of the new example. Examples in Pupils' Edition may be treated in the same way.

7. 30,142
   2
   ─────
   60,284

8. 23,103
   3
   ─────
   69,309

9. 42,301
   2
   ─────
   84,602

10. 24,130
    2
    ─────
    48,260

Show the pupils that in multiplying, as in adding, if any result is greater than 9, the left hand figure is added to the next result, which is of the same kind.

11. 63,524	12. 36,546	13. 36,426	14. 63,524
3	2	3	4
190,572	73,092	109,278	254,096

15. 26,463	16. 64,524	17. 53,625
4	4	5
105,852	258,096	268,125

18. 26,463	19. 46,035	20. 26,304
5	4	6
132,315	184,140	157,824

21. 25,036	22. 50,264
4	6
100,144	301,584

Caution: *Do not allow the pupil to write anywhere what he is to add to the next product.*
See P. Ed., p. 52.

FIRST STEPS AMONG FIGURES. 105

For addition and multiplication. (8 and rev.)

```
    a   |   b   |   c   |   d   |   e   |   f   |   g
  58473 | 62584 | 73625 | 84736 | 25847 | 36258 | 47362
  34567 | 83456 | 78345 | 67834 | 56783 | 45678 | 34567
                     h   |  i
                   5847 | 362
                   8345 | 678
```

For subtraction.

```
         a              |         b            |         c
 9  13  10  14  5  | 9  13   8  12  6  | 9  11  14  10  14
 5   6   7   8  3  | 4   5   3   4  4  | 3   8   7   6   6

         d              |         e            |         f
11  15  10   6  10 | 7  11  15  12  13 | 9  12   8  11  13
 7   8   5   3   4 | 5   6   7   8   7 | 6   5   4   3   8

         g              |         h            |         i
10   7  11   8  12 |16   7  11   8 |12   9  10
 3   4   5   6   7 | 8   3   4   5 | 6   7   8
```

For division.

```
          a                |          b               |          c
24  49  24  18   8 | 40  20   6  48  21 | 42  20  32  15  56
 6   7   8   3   4 |  5   4   3   8   7 |  6   5   4   3   8

          d                |          e               |          f
 9  24  10  30  56 | 32  28  12  64  35 | 12  30  12  21  14
 3   4   5   6   7 |  8   4   3   8   7 |  6   5   4   3   7

          g                |          h           |          i
36  15  35  16  24 | 40  18  42 | 15  16  48  28
 6   5   5   4   3 |  8   6   7 |  5   8   6   7
```

106 FIRST STEPS AMONG FIGURES.

(24)	(25)	(26)	(27)	(28)	(29)
345	235	25	435	345	54
234	543	54	543	523	43
532	455	43	234	454	25
424	345	55	455	325	54
352	432	42	543	543	24
443	554	21	235	454	55
345	345	54	352	535	32
532	553	35	544	234	54
314	345	54	535	555	43
3,521	3,807	383	3,876	3,968	384

(30)	(31)	(32)	(33)	(34)	(35)
545	435	45	543	231	45
434	554	4	355	545	53
55	345	23	234	354	44
53	534	54	542	434	53
224	242	45	345	543	25
532	555	33	554	355	44
454	343	54	435	332	23
534	543	45	543	435	54
325	254	54	343	543	45
43	442	33	555	345	53
3,199	4,247	390	4,449	4,117	439

See P. Ed., p. 59.

The following tables are an excellent preparation for short division.

Before the pupils solve the examples on page 120, give them a review of this page.

Division with remainders.

*5's (and review.)

a	b	c	d	e
16 10 3 11	22 9 6 13	5 14 4 7	18 13 11	
3 4 2 3	4 2 4 3	2 4 3 2	4 3 2	

6's (and review.)

a	b	c	d
14 16 3 26	11 26 11 9	21 10 19 7	33 14 9 5
4 3 2 6	5 4 3 6	5 4 3 2	5 6 2 3

e	f	g	h	i
11 22 18 7	13 5 38 27	20 13 8	19 7 32	
2 4 5 4	3 2 6 5	6 2 3	4 5 6	.

1. What cost 8 dozen buttons at 7 cents a dozen?

2. What cost a pair of boots at $7 and a hat at $5?

3. How much more does a reader cost at 6 shillings than a speller at 2 shillings?

---

*These are to be recited as follows: 3 in 16, 5 times and 1 remainder, &c.

4. There were 9 birds in a flock, and a hunter killed all but 4; how many did he kill?

5. How many knives at 7 shillings each may be bought for 35 shillings?

6. A boy spent 21 cents for marbles at 3 cents each; how many marbles did he buy?

7. There are 8 pigs in one pen and 5 in another; how many in both pens?

8. A boy earned 8 cents on Monday, 7 cents on Tuesday, and 6 cents on Wednesday; how many cents did he earn in the 3 days?

9. Henry bought 8 marbles at 4 cents each; what did they cost him?

10. 13 boys were skating on a pond; during the afternoon 8 of them fell upon the ice. How many of them did not fall?

11. Fred had 15 cents, he spent 5 cents for oranges and 1 cent for candy; how many cents had he left?

12. George had a bank into which he put 7 cents, his father 8, and his sister 4; how many cents had he in his bank?

13. Charles has 6 cents and his sister has 2 cents more than he; how many cents have both?

14. Henry had $8 for Christmas and his sister half as many; how many had both?

See P. Ed., p. 56.

15. William bought an orange for 4 cents, a fig for 1 cent and some candy for 2 cents. He sold them all for 12 cents. How much did he gain?

16. A boy received 6 cents a bushel for picking hops; he earned in this way 48 cents in one day. How many bushels did he pick?

17. Carrie whispered 3 times in 1 day, for each time she whispered she had to remain after school 5 minutes; how long did she have to remain?

18. 20 cents are to be divided equally among 5 boys; how many cents should each boy receive?

19. If 6 pieces of tape cost 24 cents, how much did one piece cost?

20. If a boy earned 28 shillings in 7 days, how much did he earn in 1 day?

21. Samuel walked 28 miles in 4 days; at that rate how far would he walk in 1 day?

22. How many quarts of milk at 6 cents a quart can be bought for 36 cents?

23. If 4 gallons of molasses cost 28 shillings, what cost 1 gallon?

24. 56 cents will hire how many boys for an hour, if each boy is to have 7 cents for an hour's work?

FIRST STEPS AMONG FIGURES.

(1)	(2)	(3)	(4)	(5)	(6)	(7)
52	242	543	42	34	535	344
45	525	244	54	53	454	553
31	343	554	35	4	342	435
53	451	343	43	25	535	234
44	535	234	51	40	453	544
25	434	455	45	32	341	345
53	453	542	34	41	355	453
44	345	435	34	53	434	423
35	523	252	23	45	544	345
34	254	344	45	44	355	534
55	535	535	53	35	433	453
471	4,640	4,481	459	409	4,781	4,663

95

(11)	(12)	(13)	(14)	(15)	(16)
544	543	443	532	345	432
354	441	355	344	53	544
435	355	534	434	434	434
543	434	433	453	545	542
431	542	341	345	354	344
303	344	432	524	534	453
454	354	544	343	435	341
435	425	342	234	543	535
544	543	455	555	345	344
432	254	544	443	254	455
435	434	345	344	342	532
353	543	432	535	545	344
5,263	5,212	5,200	5,086	4,729	5,300

See P. Ed., p. 59.

Pupils read :

1. 3040321   2. 30245000   3. 463317030
4. 500030261   5. 2000032   6. 15000000
7. 320030000   8. 674346537   9. 42000
10. 400320219   11. 3605000   12. 50000018
13. 463308260   14. 75000341   15. 1010010

Read the following :

1. 67345768   2. 476347854   3. 74000037
4. 735400005   5. 900007000   6. 316000140
7. 80370000   8. 700000004   9. 7020500
10. 86000045   11. 800006000   12. 90000007
13. 8060700   14. 90430000   15. 735468371

Teach pupils to write Arabic to billions, that is including 999,999,999. Teach the pupils to numerate by periods to the right as well as to the left. Thus: units, thousands, millions; millions, thousands, units, until they are perfectly familiar with it.

Method: Suppose the number ten million ninety thousand three is to be written. Instruct the pupil to write the number of millions first with a comma after it, and that the first period at the left does not need to be filled to three places by prefixing ciphers. For the above number the pupil will write 10, at first. Teacher ask 10 what? Pupil, 10 million. Teacher: What period is next to right of millions? Pupil: thousands. Teacher: How many thousands are there (in this number)? Pupil: ninety. Teacher: Write it after the comma, and, as it fills but two places, place a cipher at the left of the 90 and a comma after the 90. The number will now be 10,090. Teacher: You have now millions and thousands; what period is next? Pupil: units. Teacher: How many units are there in this number? Pupil: three. Teacher: Place the 3 to the right and prefix two ciphers to it to fill the three places of the periods. Place a period at the right because it is the end of the number, and you have 10, 090, 003.

Teach the pupil when writing numbers at the blackboard to turn directly away from it as soon as units and the period are written, for he should be sure that the number is correct without numerating to the left.

Teach the pupils of course when there are no thousands, to write three ciphers and treat unit's period in the same way.

When teaching to write billions, trillions, etc., follow the same method.

Teach the pupils that for the word hundred, you will write on the board hun.; for thousand, th.; for million, mil.; and when you get so far, for billion, bil.; for trillion, tr.; for quadrillion, quad., etc. This method will save the teacher much labor and much space on the blackboard. Thus the teacher may write upon the board:

" Write in Arabic five mil. forty th. six."

To be written on the blackboard for pupils to bring to recitation written in Arabic:

1. Write in Arabic four th. fifteen.
2. "   "   twenty mil. three hun.
3. "   "   nine mil. forty th.

Caution : Teach pupils to put a comma only after each *period*, except the last. Thus in

NOTE.—The following diagram may assist pupils in writing numbers, but after being used a few weeks the pupils should write numbers without using it.

Millions.	Thousands.	Units.
35	058	003

The teacher may draw a diagram like the above and allow the pupils to write numbers in it, as the number 35,058 003 is placed there. Teach the pupils that the third period represents millions, and that each period is read as if it stood alone, only that its name is given.

8

three hundred seventy-five million, four hundred eight thousand seven hundred forty, the pupil may have an idea that he should put a comma for hundreds, whereas the above number should be written 374,408,740. When no name is given to a number it is supposed to be units; *e. g.*, two thousand three hundred eight. Eight here means eight units, and three hundred eight, (which has nó name given to it) is 308 in units period.

The following examples may be written upon the board and the pupils required to bring the answers to class:

1. Write in Arabic, thirty million, eight thousand, threē hundred fifty-one.

2. Write in Arabic, two hundred fifty thousand.

3. Write in Arabic, one hundred sixteen million two hundred twenty.

4. Write in Arabic, three hundred million, sixty thousand. five hundred, seven.

5. Write in Arabic, five hundred thousand.

6. Write in Arabic, one million, one thousand, one.

7. Write in Arabic, seventy million, six hundred thousand, eighty.

See P. Ed., p. 61.

8. Write in Arabic, one hundred fifty-four million, two hundred sixty-one thousand, five hundred forty-eight.

9. Write in Arabic, eight hundred million.

10. Write in Arabic, three million, three.

11. Write in Arabic, ten thousand, ten.

12. Write in words, 809271300.

13. Write in Arabic, five hundred, four million, forty.

14. Write in Arabic, ten million, ten thousand.

15. Write in Arabic, one hundred one million, one hundred one.

The Roman notation uses the following letters: I=1, V=5, X=10, L=50, C=100, D=500, M=1000.

To read a number expressed in the Roman notation :

*Rule : Add the values of the letters, observing that when a letter is followed by one of greater value than itself, the difference between the two is to be taken in making up the sum.

16. Write in Roman, three hundred forty-five.

17. Write in Roman, one hundred seventy-four.

---

* From Olney's Elements of Arithmetic.

18. Write in Roman, four hundred sixty-two.
19. Write in Roman, six hundred ninety-six.
20. Write in Roman, eight hundred ninety-nine.
21. Write in words, 245306341.
22. Write in words, 32743642.
23. Write in Roman, three hundred eighty-seven.

For rapid solving.

1. $4+3+5+2+6+4+5+6+3+5+4+3+6+6+5=?$ Ans. 67.
2. $3+5+4+6+3+6+5+2+6+5+5+4+6+6+2+4=?$ Ans. 72.
3. $5+6+6+4+3+5+4+3+6+2+6+6+5+6+2+4=?$ Ans. 73.
4. $4+3+6+5+2+6+6+5+4+3+6+5+4+6+6+5=?$ Ans. 76.
5. $5+6+3+4+6+5+6+6+5+4+3+6+6+6+5+4=?$ Ans. 80.
6. $15+6\div7\times5+3\div3\times8+3+5\div7\times4+5+3-5-6=?$ Ans. 29.
7. $19+4+5\div4\times6-6-5-6\div5\times8-4\div6+5-6\times7-4=?$ Ans. 31.
8. $7\times6+5+6-4-6-6-5\div8\times6+5+6-3-6-5+6=?$ Ans. 27.
9. $6\times8+6-2-5-6-5-4\div4\times7-5-4-3-6-6\div4=?$ Ans. 8.

10. $8+7+6+5+5+4\div5\times6-5-6-6-6-3\div8+5=?$ Ans. 7.
11. $53-6-5-6-4-5+6+5+4\div6\times5-6-5-6+5=?$ Ans. 23.
12. $47+5-6-4-5-2\div5\times8-5-6-6+3\div6+8+6=?$ Ans. 21.
13. $23-5-6-4\times6+5-6-5\div7\times5-6-6\div6\times8-6=?$ Ans. 18.
14. $16\div4\times6+5-4-6-5+6+5+6+4\div7\times8-4-4-5=?$ Ans. 27.
15. $6\times7-6-5-4-3\div3+5+6+6+4+5-6-5-6-2\div5=?$ Ans. 3.
16. $28+6+6+5+4-6-5-6-4\div4\times5-6-5\div8\times7=?$ Ans. 21.
17. $61-6-6-5-5-4-6-3-6\div5\times8-5-6\div7+6=?$ Ans. 9.
18. $34-6-5-2\div3\times5-4-6+3\div7+7+5+5+6+6=?$ Ans. 33.
19. $16+5+4+3+5-6-3\div8\times5-6+4+5+5-6-4=?$ Ans. 13.
20. $27+5+4\div6\times8-6-4-5-3-6-5-4\div3\times6-4-4-6=?$ Ans. 16.

The teacher is advised to give a few examples in subtraction each day, and with them a few in multiplication and perhaps in division also.

Subtraction.

1. 3,423,056	2. 7,352,043	3. 9,635,024
654,368	834,657	746,548
2,768,688	6,517,386	8 888,476

4. 4,320,032	5. 63,140,052	6. 8,400,314
543,054	6,572,036	50,248
3,776,978	56,568,016	8,350,066

7. 7,360,042	8. 61,420,035	9. 5,304,036
2,500,075	540,257	202,356
4,859,967	60,879,778	'5,101,680

10. 8,340.050	11. 64,230,051	12. 75,310,040
762,034	510,765	230,076
7,578,016	63,719,286	75,079,964

13. $6,343,520 - 656245 = ?$ Ans. 5,687,275.

14. $94,530,062 - 8,240,277 = ?$ Ans. 86,289,785.
See P. Ed., p. 64.

*Multiplication.

1. 648,057	2. 746,805	3. 470 685
8	7	6
5,184.456	5,227,635	2,824,110

4. 8,640 753	5. 680,574	6. 358,407
7	8	6
60,485,271	5 444,592	2,150,442

7. 685,740 × 4 = ?   Ans. 2,742,960.

8. 7,406,853 × 7 = ?   Ans. 51,847971.

The teacher should solve an example in which there are two figures in the multiplier.

9. 6,354 × 23 = ?   Ans. 146,142.
10. 53,462 × 32 = ?   Ans. 1,710,784.
11. 36,425 × 34 = ?   Ans. 1,238,450.
12. 353,625 × 43 = ?   Ans. 15,205,875.
13. 563.524 × 65 = ?   Ans. 36.629,060.
14. 350,264 × 36 = ?   Ans. 12,609.504.
15. 526,304 × 34 = ?   Ans. 17,894,336.
16. 640,536 × 64 = ?   Ans 40.994,304.
17. 4675 × 45 = ?   Ans. 210,375.

---

*The numbers used in these examples are pointed off in periods for convenience in copying to blackboard or slate.

Before taking up examples in short division review division with remainders, p. 107.

In teaching pupils short division when there are remainders during the operation, write the figures of the dividend well apart; thus, in the example $379531 \div 4$, write $\frac{4)3\ 7\ 19\ 35\ 33\ 11}{9\ 4\ 8\ 8\ 2\ \cancel{4}}$ and write the remainder before the next figure as in the example given. After solving 2 or 3 examples in this way, write the figures closely, in the usual form on the blackboard, and let a pupil divide orally, the teacher using the crayon, one pupil telling how many times it is contained and what remainder, the next stating what the next partial dividend is and how many times the divisor is contained and what remainder, etc.

Let the pupils first solve the examples without remainders, given in Pupils' Edition, and the following 4 examples:

1. $24129318 \div 3 = ?$ Ans. 8,043,106.
2. $281,683,220 \div 4 = ?$ Ans. 70,420,805.
3. $12,246,921 \div 3 = ?$ Ans. 4,082,307.
4. $322,412,836 \div 4 = ?$ Ans. 80,603,209.

With remainders.

5. $83,923 \div 3 = ?$ Ans. $27,974\frac{1}{3}$.
6. $182,539 \div 4 = ?$ Ans. $45,634\frac{3}{4}$.

See P. Ed., p. 71.

7. $1,373,224 - 3 =$? Ans. $457.741\tfrac{1}{3}$.
8. $91,354\,328 \div 4 =$? Ans. $22,838,582$.
9. $273.781 \div 5 =$? Ans. $54,756^1$.
10. $1,035,879 \div 6 =$? Ans. 172.646^3.
11. $2,683,507 \div 4 =$? Ans. $670,876^3$.
12. $3,921,278 \div 6 =$? Ans. $653,546^2$.

Method of teaching pupils to add numbers like 46 and 7.

In adding 46 and 7, ask the pupil what he should add first, and either get from him or show him that 6 and 7 are to be added first; that it makes 13, of which the right hand figure is 3, which will be the right hand figure of the sum of 46 and 7, and that 46 and 7 are 53. Persevere in this plan upon the following numbers or until the pupil in adding such numbers as 37 and 8, will say at once " the right hand (or least) figure will be 5 ; 37 and 8 are 45."

45 + 6 ?	67 + 5 ?	58 + 7 ?	65 + 8 ?
86 + 5 ?	58 + 6 ?	34 + 7 ?	57 + 5 ?
26 + 8 ?	65 + 7 ?	37 + 8 ?	78 + 5 ?
86 + 7 ?	37 + 4 ?	58 + 8 ?	63 + 7 ?
47 + 7 ?	28 + 4 ?	76 + 5 ?	53 + 8 ?
37 + 6 ?	76 + 6 ?	67 + 3 ?	38 + 7 ?
64 + 8 ?	35 + 5 ?	47 + 6 ?	28 + 3 ?

FIRST STEPS AMONG FIGURES.

43+7?	64+4?	45+8?	26+5?
47+2?	38+6?	53+3?	74+7?
45+4?	76+8?	37+5?	58+2?
53+6?	24+3?	65+7?	86+4?
27+8?	68+5?	83+2?	54+6?
65+3?	36+7?	57+4?	68+8?
43+5?	54+2?	35+6;	16+3?
47+7?	38+4?	53+8?	24+5?
45+2?	86+6?	47+3?	58+7?
43+4?	34+8?	75+5?	56+2?
	27+6?	48÷3?	

The exercise above should be most used until the pupils are perfectly familiar with the combinations given, which embrace all between 8 and 3 inclusive and some are given twice. It may first be given to the pupils in the order above, then commence in the middle of the exercise and go each way. It should not be written upon the blackboard but recited orally from the reading of the examples to the pupils.

1. A boy bought a top for 18 cents and sold it so as to gain 7 cents; what did he sell it for?

2. James's mother gave him 30 cents with which to buy oranges. At 6 cents each how many could he buy?

3. Willie said he had 4 cents ; John said he had 4 times as many ; how many had John ?

4. George had 8 sticks of candy and his sister had 7 ; how many did both have?

5. 18 ripe peaches were on a tree and a bad boy stole 7 of them ; how many were left?

6. A flock of 18 birds lit upon the ground ; a hunter shot 11 of them ; how many were left? Ans. 11.

7. How many fingers have 4 boys?

8. Charlie's mother gave him 7 cents, and his sister gave him enough to make 13 cents ; how many did his sister give him?

9. John had 10 cents, one of his sisters had 7 cents and the other had 6 ; how many cents did both the sisters have?

10. Mary bought 2 yards of calico for a doll's dress ; she gave 8 cents a yard ; how much did the dress cost?

11. A tired school teacher struck a naughty boy five times upon each hand ; how many times did she strike him?

For addition and multiplication.

9's (and review.)

a	b	c	d	e	f	g
85963	74859	6374	85963	74859	62748	59637
45678	94567	8945	67894	56789	43678	94567

See P. Ed., p. 74.

h	i	j
4859	6374	
8945	6789	

For subtraction.

a				b				c			
11	16	11	13	13	15	10	12	11	16	14	12
8	9	9	5	7	6	5	4	6	7	8	9

d				e				f			
11	17	12	14	9	15	7	12	10	16	11	13
4	8	7	6	5	9	4	5	6	8	7	6

g				h				i			
8	15	13	18	10	14	13	11	9	13	15	10
5	7	8	9	4	9	4	5	6	9	8	7

j			k		
12	14	14	12	17	9
6	5	7	8	9	4

For division.

a				b				c			
42	54	25	32	24	63	16	40	30	63	48	
7	6	5	4	8	9	4	5	6	7	8	

d				e				f			
27	28	72	35	48	20	54	12	35	24	64	
9	4	8	7	6	5	9	4	5	6	8	

g				h			i		
28	42	18	24	81	40	56	45	36	30
7	6	9	4	9	8	7	9	4	5

FIRST STEPS AMONG FIGURES.

j	k	l
18 49 32 72	18 36 56	21 36 45
6 7 8 9	2 9 8	7 6 5

(1)	(2)	(3)	(4)	(5)
36	654	245	45	456
53	343	634	56	546
46	565	453	43	323
52	654	546	64	654
64	432	635	55	65
45	543	364	6	556
34	56	655	46	343
65	435	536	55	656
395	3,682	4,068	370	3,599

(6)	(7)	(8)	(9)	(10)
			6	
65	356	654	45	256
6	645	565	56	643
35	546	634	43	345
64	356	362	36	656
23	465	556	65	562
46	321	665	26	234
52	456	342	63	545
46	566	456	45	634
35	345	563	36	563
372	4,056	4,797	421	4,438

See P. Ed., p. 77.

FIRST STEPS AMONG FIGURES.

(12)	(13)	(14)	(15)	(16)	(17)	(18)
456	652	45	546	544	45	53
563	566	54	354	654	6	64
626	345	66	665	565	55	55
356	636	36	236	666	63	46
465	563	45	565	325	34	33
556	454	56	653	456	65	56
3,022	3,216	302	3,019	3,210	268	307

(1)	(2)	(3)	(4)	(5)	(6)	(7)
354	453	24	15	546	36	26
45	343	45	344	365	56	53
533	545	34	232	653	63	66
542	434	53	544	546	25	32
324	455	44	355	53	66	65
453	44	35	243	466	54	66
545	532	33	425	232	46	26
343	341	42	544	665	35	63
452	553	54	431	536	63	66
344	434	33	554	462	56	55
533	345	21	345	355	6	43
234	25	54	423	646	63	66
543	544	35	345	355	45	35
5,245	5,048	507	4,800	5,880	614	662

See P. Ed., p. 78.

FIRST STEPS AMONG FIGURES.

The answers to the examples are at the end of the book. They are placed there so that if the teacher wishes any pupil of the class to copy the examples on the blackboard for him, it may be done without the pupil's knowing what the answer is. If the teacher prefers to have the answers with the examples he can copy them from the end of the book.

In adding long columns of figures it is well to write the sum of each column separately, as follows, so that in adding each way for proof the sum of each column may be seen at a glance.

```
   56
   34
  120
   57
  ----
69396
```

Show the pupils that when 9 is added to any number the unit figure of this sum will be one less than the unit figure of the number, thus: $9+37$ is 46, $74+9$ is 83, &c.

34+6?	75+9?	26+5?	47+8?
18+4?	47+9?	24+3?	85+6?
36+9?	67+5?	48+8?	29+4?
64+7?	45+3?	26+6?	87+9?
68+5?	59+8?	74+4?	85+7?
56+3?	47+6?	78+9?	69+5?

74+8?   35+4?   56+7?   37+3?
78+6?   59+9?   75+4?   45+8?
26+4?   57+7?   78+3?   59+6?
84+9?   65+5?   46+8?   87+4?
68+7?   49+3?

Give much practice on the above exercise; it will be of great use to the pupil in all additions.

8. 463,075 × 465 = ?   9. 640,753 × 306 = ?
10. 560,423 × 204 = ?   11. 67,052 × 234 = ?
12. 574,863 × 37 = ?   13. 867,534 × 56 = ?
14. 680,574 × 78 = ?   15. 475,806 × 406 = ?

Before taking the next examples give the pupils a thorough drill in division with remainder.

Division series with remainder. (6 and rev.)

```
      a                  b                  c
23 10 3 25 8       32 11 11  7       27 22 7 17
 4  3 2  6 6        5  4  2  3        4  5 2  3

      d                  e                  f
21  8 17 20        5 33 17 13        6 13 15 26
 6  5  4  3        2  6  5  2        4  3  6  5

              g
         15 4 9 39 12
          4 3 2  6  5
```
See P. Ed., p. 80.

FIRST STEPS AMONG FIGURES.

7's (and review) with remainder.

a	b	c
22 19 42 34	69 19 14 39	28 62 17 39
3 4 5 6	7 3 4 5	6 7 3 4
d	e	f
32 23 53 13	34 29 57 47	11 31 21 52
5 6 7 3	4 5 6 7	3 4 5 6
g	h	i
51 28 27 17	47 32 26 21	49 39 25
7 3 4 5	6 7 3 4	5 6 7

8's (and review) with remainder.

a	b	c
55 26 45 24	34 16 31 53	29 43 21 29
8 7 6 5	4 3 8 7	6 5 4 3
d	e	f
41 69 13 31	20 39 28 53	76 47 22 38
7 8 3 4	3 4 5 6	8 7 6 5
g	h	i
26 28 63 32	49 33 61 39	58 31 14 23
4 3 8 7	5 6 7 8	6 4 4 3

j	k
19 39 69	44 26 17
5 6 7	8 3 4

1. $1{,}396{,}897 \div 3 = ?$ 2. $1{,}821{,}287 \div 5 = ?$
3. $2{,}144{,}698 \div 4 = ?$ 4. $39{,}164{,}794 \div 6 = ?$
5. $22{,}538{,}618 \div 6 = ?$ 6. $27{,}622{,}523 \div 6 = ?$

7. 10,763,483 ÷ 7 = ?   8. 45,171,547 ÷ 7 = ?
9. 31,752,200 ÷ 7 = ?   10. 508,019,899 ÷ 8 = ?
11. 34,856,549 ÷ 8 = ?   12. 61,476,243 ÷ 8 = ?

Unless the pupils solve the foregoing examples readily, they should review them at once.

LONG DIVISION.

The teacher may say to the pupils that when the divisor is a large number the method of short division is too difficult, illustrating by an example.

Teach the pupils that the first step in solving an example in which the divisor is greater than 12, is to place a comma after the first figure in the divisor as in the example, 5,02)73245. As in short division we cannot divide the whole of a large dividend at once, so we cannot in long division. The next step is to find how much of the dividend we will divide at first. See if the first figure of the divisor is less than the first figure of dividend or whether it is greater. In the example given it is less, (5 being less than 7). Teach the pupils that when it is less they are to count as many figures in the left of the dividend as there are figures in the divisor and place a comma after the last figure. In the given example there are three figures in the divisor, so count 3 figures in the left of the dividend and writing a comma there, the example becomes 5,02)782,45.

Require the pupils to take these two steps, (and no more) with the following examples, first placing them in form for dividing :

$76345 \div 4321$. $5738 \div 49$.
$9458 \div 875$. $875988 \div 58841$. $748567 \div 2145$.
$57881 \div 468$. $76854 \div 68$. $8768456 \div 25864$.

Teach the pupils that if the first figure of the divisor is greater than the first figure of the dividend we count one *more* figure in the dividend than there are figures in the divisor, in order that the part we take may be large enough to contain the divisor.* In the example 687531 $\div 7342$ the first figure of the divisor, 7, being greater than the first figure of the dividend, 6, we count one more figure in the dividend than the 4 figures there are in the divisor and the example with these two steps taken becomes 7,342)68753,1. Require the pupils to take these 2 steps with the following examples and

* See P. Ed., p. 98.

more if they are needed to make the entire class familiar with these steps:

$$5728{,}456 - 71235.$$
$$423214 \div 5347.$$
$$342165 \div 2543.$$
$$6847 \div 71.$$
$$5875643 \div 643.$$
$$68475032 \div 934674.$$
$$47325684 \div 3145.$$

The next step is to count the number of figures at the right of the comma in the divisor and count the same number of figures at the left of the comma in the dividend, and place a comma before the one counted which is farthest to the left: thus, in the example $34276 \div 404$, the first step is $4{,}03)34276$; the second, $4{,}03)3427{,}6$; the third, $4{,}03)34{,}27{,}6$.

Require the pupils to take these steps (and no more) with the following examples:

$$674532 \div 5342.$$
$$75694857 \div 845321.$$
$$546327 \div 643.$$
$$345367471 \div 75382.$$
$$47346 \div 23.$$
$$4326472 \div 6351.$$

8345376÷702.
56341÷68.
763542÷8547.
57345÷493.

In the example 631663÷201, which after the three steps is 2,01)6,31,663, the next step is, see how many times the number at the left of the comma in the divisor is contained in the number at the left of the first comma in the dividend. 2 is contained in 6 three times. The example becomes 2,01)6.31,663(3. Next multiply the divisor by this quotient figure, placing the first figure of the product under the figure before the last comma, thus:

$$2,01)6,31,663(3$$
$$\underline{603}$$

Next step see if you can subtract. (Teach the pupils to look at the left hand of the numbers to see whether they can subtract. If the pupils ask what is to be done when you cannot subtract, tell them you will show them in the first case in which they cannot subtract, which will not occur in the examples given for some time.) Next subtract. (Show the pupils that the remainder should be less than the divisor.)

Next see that the remainder is smaller than the divisor.

(Do not show the pupils what to do when the remainder is larger than the divisor until a case occurs in their work.) Write the next figure of the dividend at the right of the remainder.

$$2,01 \overline{)6,31,663}(3$$
$$603$$
$$\overline{286}$$

Next step count as many figures from the right of the partial dividend as there are at the right of the comma in the divisor and the example becomes

$$2,01 \overline{)6,31,663}(3$$
$$603$$
$$\overline{2,86}$$

Divide as at first and so continue the operation.

The steps are:

1st. Write the divisor and dividend in the proper form.

2d. Point off in the divisor.

3d. Place the right hand comma in the dividend.

4th. Place the left hand comma in the dividend.

5th. Divide.

6th. Multiply.

7th. See if you can subtract.

8th. Subtract.

9th. See that the remainder is less than the divisor.

10th. Write the next figure of the dividend.

11th. Point off.

Repeat steps 5, 6, 7, 8, 9, 10 and 11 until the example is solved.

The teacher should solve the following three examples *with* the pupils, before any are given them to solve alone. The teacher taking the crayon, the pupils will tell what is to be done, one pupil describing the first step, another the second and so on, or better yet, one *of them* take one of the steps then another pupil take another and so on. First solve twice the example given in the foregoing illustration.

* For method :

---

* The pupils should erase the commas which divide the number into periods before pointing off, that there be no confusion. The commas for the operation of dividing may be placed above the number instead of beneath it, if preferred.

1. $5,450,204 \div 403 = ?$
2. $162,479,845 \div 3042 = ?$
3. $20,913,844 \div 604 = ?$

Do not give the pupils more than one or two examples each day until you are sure they understand the method.

4. $4,920,352 \div 2,023 = ?$ Ans. $2,432^{416}$
5. $16,312,418 \div 5,013 = ?$ Ans. $3,254^{136}$
6. $273,785,577 \div 60,345 = ?$
7. $26,308,025 \div 4,063 = ?$
8. $189,771,597 \div 50,364 = ?$
9. $524,601,734 \div 6,047 = ?$
10. $391,838,602 \div 80,675 = ?$
11. $619,307,367 \div 80,597 = ?$
12. $278,696,736 \div 6,075 = ?$
13. $45 + 363 + 456 + 542 + 6 + 356 + 663 + 454 + 32 + 46 + 553 + 636 + 45 = ?$
14. $465 + 564 + 32 + 646 + 553 + 465 + 566 + 632 + 665 + 356 + 43 + 655 + 6 = ?$
15. $426 + 563 + 365 + 634 + 545 + 643 + 356 + 266 + 633 + 56 + 445 + 54 + 63 = ?$

1. A fox caught 5 geese which were 1-3 of the farmer's flock; how many geese in the flock?

2. A hen had 15 chickens; a cat caught 4 of them and a hawk 3. How many were left?

See P. Ed., p. 82.

3. Arthur was paid 14 cents for doing an errand; he lost 5 cents and his sister gave him 7. How many cents had he then?

4. A squirrel carried 4 nuts home one day; 5 the next day and on the third enough to make his number 16. How many did he carry home the third day?

5. James has 18 apples to divide equally among 3 boys; how many shall he give to each?

6. How many yards of tape at 2 cents a yard can I buy for 15 cents and have 1 cent left?

7. Fred gave 5 cents for an orange, 18 cents for figs and 4 cents for a lead pencil. How many cents did he spend?

8. In an orchard the trees were set 16 in a row; 7 in each row died. How many living trees in each row?

9. A boy sold a pair of doves for 25 cents and bought as many marbles at 3 cents apiece as he could for the money. How many marbles did he get and how many cents left?

10. A boy earned 9 cents one day and 12 cents the next. How much more did he earn the second day than the first?

11. Frank had 16 rabbits; he sold 3 to one boy and 4 to another. How many did he keep?

12. How long will it take a miller to grind 42 bushels of grain if he grinds 6 bushels in an hour?

13. Willie bought 2 pass-books at 5 cents each; a lead pencil for 6 cents, and 3 oranges at 4 cents a piece. What did he pay for all?

14. Willie keeps rabbits to sell; he has 20 and has 4 little houses for them. How many does he keep in each house?

15. If 4 bags contain 8 bushels of grain, how many bushels will 9 bags hold?

16. If 5 cords of wood cost $30, what will 3 cords cost?

17. If it cost 15 cents to ride 5 miles on the cars, how much will it cost to ride 7 miles?

18. How many bushels of oats will 3 horses eat in a week, if 6 horses eat 42 bushels in a week?

19. If 6 brooms cost 18 shillings, what will 5 brooms cost?

20. If a barrel of flour will last 2 men 6 months, how long will it last 1 man?

21. If 2 men consume 6 barrels of flour in a

See P. Ed., p. 86.

certain time, how much will 1 man consume in the same time?

22. If 4 horses eat 12 bushels of oats in 3 days, how many bushels will 1 horse eat in the same time?

23. If 3 teams will plow a certain field in 6 days, in how many days will 1 team plow it?

24. If 4 men can dig a certain ditch in 8 days, how long will it take 2 men to dig it?

25. If 3 men cut 6 cords of wood in a day, how much will 9 men cut in a day?

26. If it take 3 men 6 days to cut a pile of wood, how long will it take 9 men?

27. A man lost $6 by selling a cow for $37; what did the cow cost him?

28. A boy sold 4 pencils at 2 cents each, and 3 marbles at 3 cents each; how much money should he receive?

29. What is the wheat in 7 bags worth at $2 a bushel, if there are 2 bushels in each bag?

Read the following:
1. 70000580030.
2. 68000050000.
3. 680507415371.
4. 756847597547.
5. 76500000068.
6. 67459800000.

Teach pupils to write billion.

1. Write in Arabic, seventy million three thousand forty.

2. Write in Arabic, five billion four million nineteen.

3. Write in Arabic, ten billion three hundred million fifty thousand.

4. Write in Arabic, fifteen billion nine thousand.

5. Write in Arabic, two hundred billion ten million twenty.

6. Write in Arabic, forty million twenty thousand.

7. Write in Arabic, nine hundred forty billion one hundred six thousand five hundred.

8. Write in Arabic, sixteen billion sixteen.

9. Write in Arabic, five billion forty million.

10. Write in Arabic, nine hundred billion nine.

11. Write in Arabic, eight billion ninety thousand four.

Teach Roman to 1880.

12. Write in Roman, one thousand three hundred forty-one.

13. Write in Roman, nine hundred seventy-six.

See P. Ed., p. 88.

14. Write in Arabic and in words MDCCL-XXV.
15. Write in Arabic, ten billion.
16. Write in Roman, 1876.
17. Write in Roman, nine hundred forty-nine.
18. Write in words 761308260017.
20. Write 6 units of the 8th order, 8 of the 5th, 3 of the 3d and 5 of the 1st.
21. Express the following number by naming the units and their order, beginning at the left: 70900048010.

For practice in subtraction, to be given to the pupils *orally* and recited as the series have been.

49	(To be read, 45 from 49?)	73	91	53
45		64	89	48

45	69	52	38	62	35	66	19	51	22
37	66	45	34	55	29	65	14	43	19

45	82	56	32	59	36	51	85	31	56
38	66	46	28	53	29	46	66	25	54

38	73	18	30	72	46	47	63	30	57
33	67	16	25	64	43	38	59	23	54

43	50	32	64	48	61	67	48	54	39
25	48	27	56	45	54	63	39	48	37

31	70	34	50	24	51	67	65	84	56
26	64	25	46	17	48	59	63	79	48
				30	93				
				27	86				

For rapid solving.   (To be read.)

1. $3+5+9+7+8 \div 4-6 \times 9+9+5+7+6-3 \div 7 \times 9-8-8-3 \div 7+8+5 \div 3 \times 8+9-4-5-7 = ?$ Ans. 41.

2. $65-7-5-5-3 \div 9 \times 7-9-7-4 \div 5 \times 9 \div 9-8 \div 7 \times 9+8-6-2 \div 6 \times 7-5-5 \div 4 \times 8 = ?$ Ans. 64.

3. $61-5-9-5-7+9+1 \div 9+7 \div 3 \times 9+6-7-8 \div 3 \times 8-9-7-9-5 \div 7 \times 4+7+4-6 = ?$ Ans. 29.

4. $9 \times 8-6-5-7 \div 9 \times 4 \div 8 \times 6+6 \div 4 \times 7+3 \div 9 \times 6-2 \div 4 \times 6-7 \div 7 \times 8+2 \div 6 \times 4+4 \div 8 = ?$ Ans. 4.

5. $4 \times 3+8 \div 4 \times 7-3 \div 8 \times 5+4 \div 6 \times 7+8 \div 9 \times 6+6 \div 5 \times 3+2+4+2+6+8 \div 5+9 = ?$ Ans. 17.

6. $5 \times 7+7 \div 6 \times 4+7 \div 7 \times 6-4-5 \div 3 \times 4+4 \div 8 \times 5+6+6 \div 4 \times 6-3 \div 5 \times 3+8 \div 5+9 = ?$ Ans. 16.

7. $5 \times 7-3 \div 8 \times 6+6 \div 5 \times 8-3 \div 9 \times 6-6 \div 4 \times 7-6-8 \div 7 \times 5+6-1 \div 5+3 \times 7-2 \div 6 = ?$ Ans. 9.

8. $4 \times 6 + 8 \div 8 \times 9 - 4 \div 8 \times 9 - 9 \div 3 + 7 \div 4 = ?$ Ans. 4.

9. $6 \times 3 + 7 + 4 + 7 \div 9 \times 4 + 5 \div 3 \times 6 - 6 \div 6 \times 8 + 6 \div 9 \times 7 - 8 - 6 \div 4 \times 3 + 3 \div 8 \times 9 - 2 \div 5 \times 6 - 7 - 7 = ?$ Ans. 16.

10. $4 + 7 + 6 + 9 + 8 - 7 \div 3 + 8 + 8 \div 5 \times 6 - 6 - 8 \div 4 \times 8 - 7 - 6 - 7 \div 3 \times 9 + 9 \div 5 \times 9 - 7 - 8 = ?$ Ans. 66.

Problems for the slate involving Addition, Subtraction and Multiplication :

1. There are 320 rods in a mile ; how many rods in 79 miles?

2. * William has 75 cents and Charles has 68 cents more than William ; how many cents have both boys?

3. David had 123 cents ; he spent 47 cents for a ball and 39 cents for marbles. How many cents had he left?

4. The larger of two numbers is 916 and their difference is 43 ; what is the less number?

5. If the drive wheel of a locomotive turn around 352 times in going 1 mile, how many times will it turn around in going from Canandaigua to Rochester, the distance being 29 miles?

---

* In the different steps of such an example it is important as a help to mark each result.
See P. Ed., p. 90.

6. A farmer having 239 sheep, sold 99 of them and then bought 113 ; how many had he then ?

7. There are 5280 feet in a mile ; how many feet in 357 miles?

1. $374+645+57+767+436+543+675+7+454+765+577+456=$?
2. $45+776+567+457+734+475+67+754+5+676+547+375=$?
3. $6+347+75+657+746+773+46+457+347+675+766+577=$?
4. $576+745+457+674+77+556+473+765+7+657+564+705+76=$?
5. $546+375+657+74+765+257+6+75+743+656+777+467+762=$?
6. $647+756+475+367+636+753+77+345+676+45+576+767+654+365=$?
7. $6,314,532-521,987=$?
8. $463,524-39,643=$?
9. $653,425-64,287=$?
10. $475,067-36,543=$?
11. $688,045-95,387=$?
12. $4,760,352-376,534=$?
13. $4,630,024-921,045=$?
14. $34,000,435-2,700,518$?

See P. Ed., p. 92.

15. 4,500,375 − 760,187 = ?
16. 354,000,253 − 272,102,437 = ?
17. 530,024 − 543,052 = ?

1. 89,756 × 96 = ?
2. 364,758 × 356 = ?
3. 638,497 × 68 = ?
4. 498,675 × 97 = ?
5. 60,847 × 708 = ?
6. 796,805 × 705 = ?
7. 807,009 × 608 = ?
8. 6,859 × 748 = ?
9. 94,786 × 7,968 = ?
10. 603,405 − 612,134 = ?
11. 6,537,065 − 743,987 = ?

Division series with remainder.
9's and review.

```
        a                    b
32  15  68  39  61   33  49  26  46  77
 5   4   9   8   7    6   5   4   8   9
        c                    d
19  37  21  46  78   52  53  27  39  61
 4   5   6   7   8    9   6   5   4   9
        e                    f
28  54  29  34  61   35  58  41  70  43
 8   7   6   7   8    9   6   7   8   9
```

g
29 18 40 68 43
 4 5 6 7 5

h
21 88 54 25
 4 9 8 7

i
47 23 34
 6 5 4

1. 33,256,023 ÷ 7 = ?
2. 39,044,761 ÷ 6 = ?
3. 34,352,839 ÷ 5 = ?
4. 27,832,074 ÷ 6 = ?
5. 445,941,095 ÷ 7 = ?
6. 27,516,279 ÷ 4 = ?
7. 449,181,483 ÷ 6 = ?
8. 487,959,992 ÷ 7 = ?
9. 28,556,208 ÷ 6 = ?
10. 15,228,723 ÷ 4,058 = ?
11. 251,776,292 ÷ 70,486 = ?
12. 386,124 633 ÷ 60,578 = ?
13. 41,847,116 ÷ 9,048 = ?
14. 520,613,471 ÷ 70,697 = ?

Teach the pupils that when any partial dividend (after writing the next figure of the dividend at the right of the remainder) is less than the divisor, they must write a cipher in the quotient, just as they do in short division. Next

See P. Ed., p. 96.

erase the comma made in pointing off the partial dividend, write the next figure of the dividend at the right of the partial dividend, point off and proceed as before.

15. $276,265,200 \div 4,036 = ?$
16. $42,865,597 \div 6,075 = ?$
17. $397,706,673 \div 6,053 = ?$
18. $44,153,419,619 \div 70,386 = ?$
19. $15,086,456 \div 4,023 = ?$
20. $4,437,512,234 \div 60,289 = ?$

In the preceding examples the second figure from the left of the divisor has in each case been a cipher, and the examples have been so constructed that the divisor is contained as many times in each partial dividend as it appears to be. A new difficulty will arise in the following examples since the second figure of the divisor is a significant figure.

Show the pupils that when the second figure has value, the divisor is often not contained in the partial dividend as many times as it appears to be, since in multiplying the divisor by the quotient figure there will usually be something to add to the product of multiplying the first figure, coming from the product of multiplying the second. Teach the pupils to observe how much there will be to add to this first product and to allow for it.

Teach them also that when any product is greater than the partial dividend, it shows that the quotient figure which gave that product is too large, and that the partial product and that quotient figure must be erased and a smaller quotient figure used.

Use the following examples in illustration to the class, or at least as many examples as will make the matter clear to the class:

$267,142 \div 352 = ?$
$2,659,478 \div 461 = ?$

Use too large a quotient figure in some instances so as to show the pupils what to do when they use one that is too large. Teach them to use much care in finding the quotient figure and so save themselves much work. Show that the divisor never should be contained in the partial dividend 10 times.

1. $4,367,695 \div 673 = ?$
2. $220,396 \div 254 = ?$
3. $307,627 \div 354 = ?$
4. $4,925.151 \div 694 = ?$
5. $6,310,318 \div 781 = ?$
6. $3,280,381 \div 482 = ?$

See P. Ed., p. 98.

FIRST STEPS AMONG FIGURES. 149

7. $239,294,268 \div 3,642 = ?$
8. $334,990,637 \div 574 = ?$

For rapid solving. (To be read and to be answered without use of slate.)

1. $6 \times 6 - 8 \div 4 \times 6 + 8 - 2 \div 6 \times 4 + 8 \div 5 \times 9 - 9 \div 9 \times 6 + 8 + 9 = ?$ Ans. 59.
2. $9 \times 6 - 5 \div 7 \times 6 - 6 \div 4 \times 7 + 5 + 4 \div 9 + 9 + 6 + 8 - 4 \div 3 = ?$ Ans. 9.
3. $5 \times 8 - 6 - 7 \div 3 \times 5 + 7 + 6 + 5 \div 9 \times 6 + 3 \div 5 \times 3 + 9 - 7 = ?$ Ans. 29.
4. $8 \times 7 - 9 - 5 \div 6 \times 3 - 3 \div 3 \times 7 + 3 \div 5 \times 6 - 6 - 6 \div 6 \times 4 = ?$ Ans. 28.
5. $6 \times 8 - 6 - 7 \div 5 \times 8 + 4 + 3 \div 7 \times 4 + 7 + 5 \div 8 + 7 + 5 + 9 = ?$ Ans. 27.
6. $4 \times 9 + 8 + 4 \div 6 \times 4 - 5 \div 3 \times 6 - 8 - 4 \div 6 \times 5 + 8 + 4 = ?$ Ans. 47.
7. $7 \times 6 - 5 - 9 \div 4 \times 8 - 8 - 6 - 2 \div 8 \times 5 + 9 + 7 - 8 = ?$ Ans. 33.
8. $6 \times 4 + 8 \div 8 \times 9 + 6 \div 7 \times 8 + 1 \div 7 \times 9 - 4 - 3 \div 7 + 9 = ?$ Ans. 17.
9. $9 \times 8 - 6 - 9 - 1 \div 7 \times 5 + 5 \div 9 \times 6 + 8 + 4 \div 6 \times 8 + 7 \div 9 \times 4 + 6 + 9 + 5 \div 8 \times 6 + 7 + 2 \div 9 \times 8 - 8 - 5 \div 9 = ?$ Ans. 3.
10. $7 \times 6 - 6 + 9 \div 5 \times 6 - 8 - 9 + 5 \div 7 \times 8 - 6 - 5 - 5 \div 8 \times 9 + 9 - 7 + 6 + 9 + 3 \div 8$

$\times 4+9+8\div 9\times 7+8+5\div 8\times 7 = ?$
Ans. 42.

11. $9\times 6-8-7+6\div 9\times 6-6\div 3\times 7+8-7$
$+6\div 7\times 4+9+8-7-4\div 6\times 8-7$
$-6-7\div 4+7+8+8\div 4\times 7+8 = ?$
Ans. 64.

12. $8\times 8+6+3-7-8-4\div 6\times 7-8-6\div 7$
$\times 8-8\div 6\times 9+3-8-4\div 9+8+9+$
$7+6+9-7+5+8-7 = ?$ Ans. 45.

13. $6\times 8+3-7-8\div 9\times 7+9+8\div 5\times 6+$
$2\div 7\times 9-7-9-8\div 8\times 7-8-7\div 3$
$+7+8+7 = ?$ Ans. 31.

14. $7\times 9-8-7\div 8\times 9+8-5-6-4-5\div$
$6\times 8+9+7\div 9+9+9+8+2\div 4\times 7$
$-7\div 8\times 5+8+9+8 = ?$ Ans. 60.

1. How far will a boy walk in 7 days, walking 9 miles each day?

2. How far will a boy walk in 2 days, walking 9 miles the first day and 7 miles the second day?

3. In how many days will a man earn 48 shillings, at 8 shillings a day?

4. A boy earned 45 cents Monday, and 53 cents Tuesday; how much more did he earn Tuesday than Monday?

See P. Ed., p. 99.

5. If 4 peaches cost 8 cents, what cost 9 peaches?

6. If a merchant sells 9 spools of thread in 3 hours, at that rate how many would he sell in 1 hour?

7. If 3 girls can make 6 aprons in a day, how many can one girl make in a day?

8. If 2 girls can do a piece of work in 8 days, in how many days can 1 girl do it?

9. A boy bought 9 marbles and lost all of them but 3; how many did he lose?

10. A farmer bought a pig for $6 and sold it for $9; how much did he gain?

11. There were 8 cows in a field and 6 more were put in; how many were in the field then?

12. There are 17 girls in a class and 9 boys; how many pupils in the class?

13. There are 16 caps in the entry and 7 bonnets; how many more caps than bonnets in the entry?

14. If John is well how many days should he come to school in 4 weeks?

15. A man may rightly work how many days in 3 weeks?

16. James has 5 cents and his sister has 2 cents less than twice as many; how many have both?

Examples for slate in Addition, Subtraction, Multiplication and Division :

1. A has 9 fields, containing in all 197 acres ; B has 13 fields, containing 239 acres ; C has 17 fields, containing 298 acres ; D has 6 fields, containing 85 acres ; how many fields and how many acres have all ?

2. How many horses at $185 each can be bought for 25 cows at $37 each?

3. If 69 acres of land cost $6,486 what will 207 acres cost ?

4. I borrowed of Mr. Rawson at one time $697, at another $1,748, and at another $456; I paid him $975; how much do I still owe him?

5. What is the sum of eighteen thousand three, nine million twenty thousand, eight hundred six, seven thousand sixty, 95 thousand seven hundred, twenty-one million five hundred seventy-six, and ten million ten ?

6. Henry's kite was up in the air 375 feet, it then fell 98 feet and then rose 268 feet; how high was it then?

7. Three men bought a hotel for $25,800; the first paid $6,790, the second twice as much,

See P. Ed., p. 102.

FIRST STEPS AMONG FIGURES. 153

and the third the remainder ; how much did the third pay ?

8. The earnings of a father and his 3 sons for a year amount to $2175 ; their expenses are $957 ; if the balance is divided equally, how much will each have ?

9. If four dresses of 15 yards each are cut from 78 yards of calico, how many yards will be left ?

1. 7 + 46 + 74 + 35 + 47 + 73 + 64 + 57 + 75 + 6 + 77 + 45 + 63 + 57 + 46 + 75 = ?
2. 376 + 455 + 757 + 463 + 375 + 747 + 654 + 37 + 576 + 346 + 775 + 7 + 464 + 647 + 356 + 565 = ?
3. 475 + 647 + 756 + 765 + 437 + 674 + 575 + 756 + 647 + 567 + 456 + 743 + 357 + 556 + 463 + 756 = ?
4. 76 + 35 + 8 + 47 + 85 + 37 + 56 + 87 + 75 + 84 + 68 = ?
5. 58 + 765 − 485 + 678 + 537 + 6 + 753 + 488 + 846 + 587 + 758 = ?
6. 678 + 845 + 784 + 326 + 487 + 856 + 678 + 588 + 865 + 478 + 756 = ?
7. 7 + 58 + 84 + 56 + 78 + 64 + 87 + 6 + 58 + 75 + 86 + 68 + 75 = ?

8. $648+785+874+688+576+845+688+786+75+847+687+58+6=$?
9. $758+875+684+768+475+886+744+358+652+887+546+785+648=$?
10. $8+57+68+84+75+7+58+76+88+47+63+78+86+55+67+88=$?
11. $368+475+638+857+583+646+878\ 57+645+768+582+7+676+848\ 587+766=$?
12. $67+788+856+475+687+878+564+87+656+478+880+567+375+688\ 856+785=$?
13. $7,963,034-546,573=$?
14. $758,600.341-79,423,275=$?
15. $8,460,075-987,286=$?
16. $658,000,468-35,030,273=$?
17. $43,750,078-44,345,621=$? Impossible
18. $6,475,000,374-293,030,596=$?

1. $97,806 \times 59=$?
2. $97,865 \times 896=$?
3. $96,897 \times 6,978=$?
4. $5,740 \times 7,500=$?
5. $746,800 \times 9,000=$?
6. $470,900 \times 70,580=$?

See P. Ed., p. 110.

FIRST STEPS AMONG FIGURES. 155

7. 869,070 × 670,900 = ?
8. 790,600 × 806,700 = ?
9. 62,802,889 ÷ 9 = ?
10. 71,262,955 ÷ 8 = ?
11. 538,908,792 ÷ 7 = ?
12. 376,571,086 ÷ 7 = ?
13. 339,253,657 ÷ 9 = ?
14. 324,763,528 ÷ 7 = ?
15. 446,217,169 ÷ 9 = ?
16. 54,284,406 ÷ 8,047 = ?
17. 25,534,849 ÷ 7,197 = ?
18. 45,126,612 ÷ 914 = ?
19. 4,368,565 ÷ 9,168 ?
20. 63,008,141 ÷ 5,274 = ?
21. 47,230,943 ÷ 79 = ?
22. 9,290,055,741 ÷ 4,869 = ?
23. 33,279,851 ÷ 48,600 ?
24. 68,643,216 ÷ 87,000 ?
25. 76,845,678 ÷ 100 = ?
26. 3,921,534,261 ÷ 486,000 = ?
27. 60,064,175 ÷ 8,000 = ?
28. 90,700 × 50,700 = ?
29. Subtract 3 billion 6 thousand 750 from 45 billion 1 million seven hundred sixty-three thousand 4 hundred.
30. 284,553,437 ÷ 3,790 = ?

See P. Ed., p. 116.

31. 4,167,300,326 ÷ 4,790,000 = ?
32. 5,074,000 × 68,070 = ?
33. 3,424,330,021 ÷ 497,000 = ?
34. 2,468,576,216 ÷ 10,000 = ?

12's (and review.)
For addition and multiplication.

```
         a                    b                      c
    9  12   8  10        7  11   9  12          8  10   7  11
    8   9  10  11       12   8   9  10         11  12   8   9
         d                    e                      f
    3  12   8  12        7  11   9  12          8  10   7  11
   12  11  12   4        9  10  11  12          8   9  10  11
         g                    h                      i
   12   5  12          9  12   8  10   7  11       6  12   4
    3  12   6         12   8   9  10  11  12      12   5  12
```

For subtraction.

```
           a                              b
  22  19  22  18  19  19         21  18  21  17  20  18
  12  11  10   9   8  12         11  10   9   8  12   8
           c                              d
  16  21  23  19  20  15         20  24  16  19  17  22
   9  10  11  10   9   8         11  12   8   9  10  11
                        e
                21  20  17  20  23  18
                12   8   9  10  12  11
```

See P. Ed., p. 140.

FIRST STEPS AMONG FIGURES. 157

For division.

```
       a                    |        b
72 121 108 96 72 100       |132 77 80 110 84 48
12  11  12  8  9  10       | 12 11 10  11 12 12

       c                    |        d
120 108 72 132 90 99        | 56 88 120 81 96 80
 12   9  8  11 10  9        |  8 11  10  9 12  8

     e                      |        f
   60 36                    | 63 110 99 144 64 90
   12 12                    |  9  10 11  12  8  9
```

1. 799,896 × 12 = ?
2. 6,347,435 × 12 = ?
3. 7,968,473 × 11 = ?
4. 3,546,247 × 12 = ?
5. 989,769 × 12 = ?
6. 799,958 × 12 = ?
7. 8,989,978 × 12 = ?

12's (and review.)
Division with remainders.

```
          a                 |         b
127 76 94 70 150 108        |115 71 96 137 106 85
 11 10  9  8  12  11        | 10  9 10  11  12  8
         ·c                 |         d
116 102 78 105 86 139       |124 88 94 95 117 89
 12   8  9  10 11  12       | 10  9  8 12  11 10
```

See P. Ed., p. 117.

e
115 78 96 130 63 107
 9 8 11 12 8   9

For more practice.

For division.

a
86 152 40 68 47
11 12 6 7 8

b
78 130 94 62 34
 9 11 12 7 6

c
118 75 114 61 83
 12 11   9 8 7

d
52 55 88 64 107
 6 8 9 11 12

e
69 54 102 89 45
 6 7   8 7 6

f
142 86 52 78 48
 12 11 9 8 7

g
76 93 70 139 81
 6 8 9 11 12

h
57 41 70 107
 6 7 8   9

8. $9{,}587{,}888{,}171 \div 12 = ?$
9. $95{,}621{,}647 \div 12 = ?$
10. $813{,}764{,}564 \div 11 = ?$
11. $83{,}645{,}840 \div 12 = ?$

See P. Ed., p. 118.

FIRST STEPS AMONG FIGURES.   159

For rapid solving. (To be read and to be answered without the use of slate.)

1. $17+4 \div 3 \times 9+8+7+6 \div 12 \times 9-8-6+7 \div 7 \times 6+9+6 \div 9 \times 11+9+8+7+7 \div 9 \times 10 \div 3+9+9 \div 2+7 \div 2 = ?$ Ans. 18.

2. $144 \div 2 \div 2 \div 9 \times 8 \times 2-1 \div 7 \times 12 \div 3+6 \div 6 \times 12 \times 2 \div 3 \div 2 \times 3+9+9 \div 3 \div 2+6+7 \times 4 \div 10 \times 8 \div 4 \times 3 \div 8 = ?$ Ans. 9.

3. $9 \times 12 \div 2 \div 3 \times 2-4 \div 2 \times 3 \div 12+7 \times 12 \div 3 \times 2 \div 4+12 \div 2 \times 3+6 \div 3 \times 2+7 \div 5 \times 8 \div 3 \times 2 \div 3 \times 4 \div 8+9+8+9 \times 3+6 \div 9 = ?$ Ans. 12.

4. $19 \times 2+4 \div 7 \times 12 \div 2 \times 3 \div 12 \times 11 \div 3 \times 4 \div 12 \times 8+8 \div 2+3 \div 3 \times 2+8 \div 6 \times 12 \div 3 \times 2-2 \div 3 \times 2+12 \div 3 \times 5 \div 4 \times 10 \div 4 \div 2 \times 4-4 \div 12 = ?$ Ans. 8.

5. $13 \times 3+8+7 \div 6 \times 8 \div 3 \times 2 \div 3 \times 2+7+7+6 \div 2-10 \times 3-8 \div 5+7 \times 4 \times 2 \div 4 \div 15 \times 9 \times 2 \times 3-9 \div 3 \times 4 \div 11 \times 8 \div 3 = ?$ Ans. 32.

6. $14 \times 2 \div 4 \times 8 \div 2 \times 3 \div 7 \times 8 \div 3 \div 2 \times 4+8 \div 2 \times 3 \div 2 \div 3 \times 2 \times 2 \div 12 \times 9+8+9-6+10+3 \div 2+9 \div 3 \times 2 \div 4 = ?$ Ans. 8.

7. $12 \times 11 \div 3 + 8 \div 2 + 8 \div 2 \times 3 - 3 \div 2 \div 4$
$\times 9 + 3 \div 3 \times 4 \div 2 + 8 \div 2 \times 3 + 3 \div 12$
$\times 9 \div 3 - 2 \div 4 \times 9 \div 3 \times 2 \div 3 \times 2 \div 8$
$\times 4 \times 2 = ?$ Ans. 16.

8. $19 \times 2 + 4 \div 3 \times 2 \div 4 \times 12 + 8 \div 2 \div 2 + 5$
$\div 2 \times 3 \div 2 \times 4 \div 12 \times 8 \div 2 \times 3 + 4 \div 8$
$\times 9 + 1 \div 4 \times 3 \div 5 \times 2 + 8 \div 2 - 4 \times 4$
$= ?$ Ans. 60.

9. $9 \times 3 - 5 \div 2 \times 3 + 3 \div 2 \div 9 \times 18 + 9 + 8$
$+ 5 \div 2 \times 3 + 9 \div 8 \times 3 \div 2 \times 3 \div 2 \times 3$
$\div 9 \times 4 \div 2 \times 5 \div 2 \div 5 \times 6 \div 3 = ?$
Ans. 18.

10. $16 \times 4 \div 2 \div 2 \times 3 + 12 \div 4 \times 3 \div 5 \times 4 \div 2$
$\times 4 \div 2 - 6 \div 2 \times 5 \div 25 \times 12 + 8 + 12$
$\div 2 \div 4 \times 10 \div 2 \div 7 \times 15 - 50 \times 3 \times 2$
$\div 3 - 25 \div 5 = ?$ Ans. 5.

1. If 4 lemons cost 7 cents, what cost 20 lemons?

Explanation: Teach the pupils that if a certain quantity of anything cost a certain amount, 3 times that quantity will cost 3 times as much; 4 times that quantity will cost 4 times as much, etc.

Solution: If 4 lemons cost 7 cents, 20 lemons, which are 5 times 4 lemons, will cost 5 times 7 cents, or 35 cents.

See P. Ed., p. 119.

2. If 3 oranges cost 10 cents, how many oranges may be bought for 30 cents?

Explanation : Teach the pupils that if a certain sum of money will buy a certain quantity, 3 times that sum will buy 3 times that quantity, etc.

Solution : If 10 cents will buy 3 oranges, for 30 cents, which are 3 times 10 cents, you can buy 3 times 3 oranges, or 9 oranges.

3. What cost 18 spools of thread at the rate of 2 spools for 9 cents?

4. If 2 knives may be bought for 5 shillings, what will 20 knives cost?

5. If 2 men cut 5 cords of wood in a day, how many cords will 10 men cut in a day?

6. If 3 bushels of wheat cost $6, what will 8 bushels cost?

7. If 2 bushels of wheat cost $3, how many bushels may be bought for $18?

8. 36 cents will buy how many marbles at 3 for 4 cents?

9. If 3 boys can do a certain work in 6 days, how many days will it take 1 boy to do the same work?

10. If 2 men can hoe a field of corn in 4 days, how many days will it take 1 man to do it?

11. If 3 men can cradle 6 acres of grain in a day, how many acres can 1 man cradle in a day?

12. If 2 men can build a wall in 6 days, how many men can build it in 1 day?

13. If 4 men can dig a ditch in 12 days, how many men can dig it in 1 day?

14. If 2 men can dig 8 rods of ditch in 1 day, how many rods can 1 man dig in a day?

15. If 3 men can dig a ditch in 12 days, how many days will it take 4 men?

Call attention of pupils to the difference between the 15th example and the 16th, and teach them to find about 1 of the kind the question asks about. For instance the 15th asks about 4 men, hence find out how many days it will take 1 man. In the 16th it asks about 6 days, hence find about 1·day?

Solution of 15th: If 3 men can dig it in 12 days it will take 1 man 3 times 12 days, or 36 days; and *four* men can dig it in ¼ of 12 days, or 3 days.

16. If 3 men can dig a ditch in 12 days, how many men can dig it in 6 days?

Solution: If 3 men can dig it in 12 days, to dig it in 1 day, it will take 12 times 3 men or

See P. Ed., p. 120.

36 men ; and to dig it in 6 days it will take 1-6 of 36 men, or 6 men ?

17. A boy lost 4 marbles, then bought 6, and losing 10 he has 35 ; how many had he at first ?

18. What number multiplied by 3 will give 12 ?

19. What number subtracted from 7 will leave 4 ?

20. How many days will it take 8 men to do a work that requires 6 men 12 days ?

21. How many men will do a work in 25 days that takes 5 men 10 days ?

22. What cost 9 suits of clothes at $14 for each coat, $2 for each vest, and $4 for each pair of pants ?

23. How many oranges at 6 cents each can be bought for 4 cents and 5 lemons at 4 cents each ?

24. A boy gave 10 marbles worth 7 cents for 3 figs worth 2 cents each ; how much did he lose ?

25. What cost 60 eggs at 12 cents a dozen ?

26. A boy has 33 cents, how many marbles at 3 cents each can he buy and keep 6 cents ?

27. A boy has 39 cents, how many must he earn that he may buy a dozen oranges at 4 cents each ?

28. In Mary's garden are 8 roses, twice as many pinks and a dozen daisies; how many flowers in her garden?

29. If 3 pounds of sugar cost 24 cents, what will half a pound cost?

30. Mary has 8 cents, her sister has 6 cents, and their brother has half as much as both of them; how many have the three children?

31. If I buy 60 chickens at the rate of 5 for $2, and sell them at the rate of 12 for $5, how much will I gain?

32. How far apart will 2 men be in 7 hours, if they start from the same place, and travel in opposite directions, one 6 miles an hour and the other 4 miles an hour? How far if they travel in the same direction?

33. A man who drives 9 miles an hour is trying to overtake a man who is 24 miles ahead of him and who goes 6 miles an hour; in how many hours will he overtake him?

34. How many ducks at the rate of 7 for $6 can I buy for $29 and have $5 left?

The pupils should mark each answer, and also its denomination. They should be required to mark not only the *denomination* of each result in the process of solving problems, but what it represents, that is whether it is cost, selling price, gain or loss, A's number, B's number, &c. In this way they will succeed with many problems on which they would otherwise fail.

EXAMPLES FOR THE SLATE.

1. A had $8,948 to which he added $2,284, and then he lost $1,632 when he used all he had in buying 38 village lots; how much did each lot cost?

2. B bought 265 acres (of land) for $22,790; sold 169 acres of it at $97 an acre and the rest at cost. Whole gain?

3. A horse and 16 oxen are worth $1439, and the horse is worth $175; what are the oxen worth. What is each oxen worth?

4. Paid 36 barrels of flour for 60 yards of cloth at $6 a yard; how much was the flour a barrel?

5. If the front and rear walls of a house each contain 37,390 bricks, and the other two walls each 49,758; how many bricks in the four walls?

6. If 15 boys walk 900 miles in 60 days, how far will they walk in 2 days?

See P. Ed., p. 125.

7. Add forty-five million nine thousand ten, fifty thousand eight hundred, nine million nine hundred thousand seven hundred nine, ninety million ninety thousand seven, and six hundred seventy-eight.

8. A sold one horse for $185, another for $165, and another for $187 ; what was the average price of a horse?

9. Divide the product of 6580 and 7900 by their sum.

10. A bought 300 acres of western land for 1,200 ; B bought 275 acres for $175 less, and C 125 acres at $4 an acre ; how many acres did the 3 men buy? How much did they pay?

11. A grocer bought 279 pounds of butter at 27 cents a pound and 98 pounds at 26 cents a pound ; he sold the whole at 32 cents a pound ; how much did he gain?

12. The weight of a number of hogs was as follows: 250 pounds, 245 pounds, 260 pounds, 257 pounds, 273 pounds and 293 pounds ; what was their average weight?

13. A man wishes to buy a piano for $375 ; he lays up $5 a week for a year, or 52 weeks ; how much more must he save to get the piano?

14. Two men start from the same place at

See P. Ed., p. 128.

the same time and travel in the same direction, one at the rate of 35 miles a day and the other 44 miles a day; how far apart are they at the end of 4 days? How far apart if they had traveled in opposite directions?

15. Divide the product of the sum and difference of 364 and 93 by the difference between their sum and difference.

16. A farmer bought one cow for $34, another for $43 and another for $61; what was the average price of the cows?

17. The product of two numbers is 1,017,702 and one of them is 2,758; what is the other?

18. What is the sum of seventy thousand nine, nineteen thousand six hundred forty-nine, nine million seven hundred thousand, six hundred thousand nine hundred eight, fifty million sixty, and three hundred seventy-nine thousand eight hundred ninety-eight?.

19. The remainder is 713, the quotient 579, the divisor 2758; what is the dividend?

20. A man bought 5 horses at $165 each and 6 more for $902; what was the average price paid?

21. A woman left her four children $15,000; the eldest received one-half of it, and the re-

mainder was divided equally among the other children; what was the share of each?

22. A, B and C sold 20 village lots for $14,600; A received twice as much as B, and B $200 more than C, whose share was $3,500; what did A receive? B receive?

23. A company of 14 miners sell a mine in 1,245 shares at $210 per share; what does each receive?

1. $49+7,068+9,847+958+37+489+8,956+9843= ?$
2. $946+378+795+849+696+784+359+436+775+898= ?$
3. $547+397+484+758+969+847+958+497+384+947+358+596= ?$
4. $567+498+948+397+846+372+458+796+389+486+958+347+598= ?$
5. $123+456+789+987+456+321+743+398+476+395+948+767+496+324= ?$
6. $578+397+956+789+437+496+875+749+658+976+345+876+901= ?$
7. $947+643+358+895+769+576+348+954+847+659+438+987+648+326= ?$

FIRST STEPS AMONG FIGURES.

8. $756+895+468+347+579+943+658$
$+547+892+675+487+949+673$
$+246+987=?$

9. $391+849+327+496+327+843+659$
$+742+869+324+496+932+783$
$+468+579+453=?$

10. $938+493+745+679+548+987+765$
$+899+624+345+879+354+497$
$+384+947+486+849+435=?$

TEACHERS' EDITION.

PAGE.	NO.	ANSWER.
126–	1.	5,245.
	2.	5,048.
	3.	507.
	4.	4,800.
	5.	5,880
	6.	614.
	7.	662.
128–	8.	215,320,875.
	9.	196,070,418.
	10.	114,326,292.
	11.	15,090,168.
	12.	21,269,931.
	13.	48,581,904.
	14.	53,084,772.
	15.	193,177,236.
129–	1.	465,632 [1-2].
	2.	364,257 [2-5].
	3.	530,174 [2-4].
	4.	6,527,465 [4-8].
	5.	3,756,436 [2-6].
	6.	4,608,753 [5-6].
130–	7.	1,537,640 [1-7].
	8.	6,453,078 [1-7].
	9.	4,536,028 [4-7].
	10.	63,502,487 [5-6].
	11.	4,857,068 [5-6].
	12.	7,684,530 [3-9].
136–	1.	13,524 [72].
	2.	53,412 [641].
	3.	34,625 [344].
	4.	2,432 [116].
	5.	3,254 [116].
	6.	4,537 [312].

PAGE.	NO.	ANSWER.
	7.	6,475 [100].
	8.	3,768 [45].
	9.	86,754 [399].
	10.	4,857 [127].
	11.	7,684 [19].
	12.	45,876 [80].
	13.	4,197.
	14.	5,648.
	15.	5,049.
143–	1.	25,280 rds.
	2.	218 cts.
	3.	37 cts.
	4.	873.
	5.	10,208 times.
144–	6.	253 sheep.
	7.	1,884,960 ft.
	1.	5,756.
	2.	5,478.
	3.	5,472.
	4.	6,832.
	5.	6,160.
	6.	7,139.
	7.	5,792,545.
	8.	423,881.
	9.	589,138.
	10.	438,524.
	11.	592,658.
	12.	4,383,818.
	13.	3,708,979.
	14.	31,299,917.
145–	15.	3,740,188.
	16.	81,897,816.
	17.	Impossible.

1. 8,616,576.
2. 129,853,848.
3. 43,417,796.
4. 48,371,475.
5. 43,079,676.
6. 561,747,525.
7. 490,661,472.
8. 5,130,532.
9. 755,254,848.
10. Impossible.
11. 5,793.078.
146- 1. 4,750,860 [3-7].
2. 6,507,460 [1-4].
3. 6,870,567 [4-8].
4. 4,638,679.
5. 63,705,870 [4-7].
6. 6,879,069 [2-4].
7. 74,863,580 [3-4].
8. 69,708,570 [2-7].
9. 4,759,368.
10. 3,752 [3,107].
11. 3,572 [100].
12. 6,374 [461].
13. 4,625 [114].
14. 7,364 [743].
15. 68,450 [1,000].
147-16. 7,056 [897].
17. 65,704 [341].
18. 627,304 [273].
19. 3,750 [204].
20. 73,004 [678].
148- 1. 6,489 [699].
2. 867 [176].
3. 869 [1].
4. 7,096 [627].
5. 8,079 [619].
6. 6,805 [871].
149- 7. 65,704 [300].
8. 583,607 [219].
152- 1. 45 fields; 819 A.
2. 5 horses.
3. $19,458.
4. $1,926.
5. 40,142, 155.
6. 545 ft.

7. $5,430.
153- 8. $304 [3-4].
9. 18 yds.
1. 847.
2. 7,600.
3. 9,630.
4. 658.
5. 5,961.
6. 7,341.
7. 802.
154- 8. 7,563.
9. 9,066.
10. 1,005.
11. 9,381.
12. 9,687.
13. 7,416,461.
14. 679,177,066.
15. 7,472,789.
16. 622,970,195.
17. Impossible.
18. 6,181,969,778.
1. 5,770,554.
2. 87,687,040.
3. 676,147,266.
4. 43,050,000.
5. 6,721,200,000.
6. 33,236,122,000.
155- 7. 583,059,063,000.
8. 637,777,020,000.
9. 6,978,098 [7-9].
10. 8,907,869 [3-6].
11. 76,966,970 [2-7].
12. 53,795,869 [87].
13. 37,694,850 [7-9].
14. 46,894,789 [6-7].
15. 49,579,685 [4-9].
16. 6,745 [7,391].
17. 3,547 [7,000].
18. 49,372 [604].
19. 476 [4,897].
20. 11,946 [4,987].
21. 597,860 [8].
22. 1,906,000 [3,741].
23. 684 [87,461].
24. 789 [316].

FIRST STEPS AMONG FIGURES.

25. 768,456 [70].
26. 8,069 [201].
27. 7,508 [175].
28. 4,508,490,000.
29. 42,001,756,050.
30. 75,080 [237].

156-31. 870 [224].
32. 345,387,180,000.
33. 6,890 [21].
34. 246,857 [6,216].

157- 1. 9,598,752.
2. 76,169,220.
3. 87,653,203.
4. 42,554,904.
5. 11,877,228.
6. 9,599,496.
7. 107,879,736.

158- 8. 798,950,080 [11-12].
9. 7,968,470 [7-12].
10. 73,978,596 [6-11].
11. 6,970,486 [6-12].

165- 1. $252 [24-34].
2. $1,859.
3. $1,264; $79.
4. $10.
5. 174,296 bricks.
6. 30 ml.

166- 7. 145,051,904.
8. $179.
9. 3,589 [12.280-14.480].
10. 700 A.; $2,725.
11. 1,983 cts.
12. 263 lbs.
13. $115.
14. 36 ml.; 316 ml.

167-15. 605 [137-184].
16. $46.
17. 369.
18. 60,770,524.
19. 1,597,595.
20. $157 [$2,500 each.
21. Eldest $7,500; oth's

168-22. A, $7,400; B, $3,700.
23. $18,675.
1. 37,247.
2. 6,916.
3. 7,742.
4. 7,660.
5. 7,679.
6. 9,033.
7. 9,395.

169- 8. 10,102.
9. 9,538.
10. 11,854.

END OF KEY TO TEACHERS' EDITION.

PUPILS' EDITION.

PAGE.	NO.	ANSWER.	PAGE.	NO.	ANSWER.
33-	1.	251.		34.	317.
	2.	272.	37-	35.	21,752.
	3.	271.		36.	24,853.
	4.	239.		37.	63,545.
	5.	234.		38.	22,544.
	6.	239.		39.	93,733.
	7.	262.		40.	62,592.
	8.	324.		41.	66,651.
	9.	252.		42.	77,484.
	10.	260.		43.	28,345.
34-	11.	296.		44.	72,353.
	12.	295.		45.	62,156.
	13.	308.		46.	63,234.
	14.	335.		1.	46,206.
	15.	314.		2.	90,693.
	16.	325.		3.	39,606.
	17.	313.		4.	42,064.
	18.	327.	38-	5.	93,069.
	19.	324.		6.	24,640.
35-	20.	316.		7.	60,369.
	21.	317.		8.	69,396.
	22.	346.		9.	62,840.
	23.	319.		10.	96,903.
	24.	302.		11.	60,846.
	25.	345.		12.	69,369.
	26.	307.		13.	32,341.
	27.	297.		14.	31,203.
36-	28.	325.		15.	10,243.
	29.	328.		16.	20,413.
	30.	308.		17.	13,023.
	31.	293.		18.	41,302.
	32.	328.		19.	14,032.
	33.	303.		20.	30,218.

FIRST STEPS AMONG FIGURES. 175

21. 13,024.
22. 40,132.
23. 20,312.
39-24. 2,429.
25. 2,463.
26. 2,544.
27. 2,420.
28. 2,349.
29. 2,537.
30. 2,917.
31. 2,996.
32. 2,364.
40-33. 2,967.
34. 3,178
35. 3,184.
36. 34,892.
37. 36,775.
38. 35,561.
41-39. 35,673.
40. 35,680.
41. 4,655.
42. 4,514.
43. 4,615.
42-44. 4,300.
45. 4,522.
46. 4,837.
47. 40,044.
48. 48,573.
49. 48,874.
43-50. 49,083.
51. 51,204.
47- 1. 4,258.
2. 4,011.
3. 4,741.
4. 4,743.
5. 4,893.
6. 4,872.
48- 7. 15,435.
8. 65,144.
9. 36,425.
10. 653,453.
11. 752,326.
12. 42,453.
13. 42,334.
14. 8,536.

15. 563,135.
16. 741,027.
17. 411,303.
18. 623,024.
19. 71,634.
20. 44,252.
21. 443,043.
22. 476,446.
23. 556,467.
24. 375,667.
49-25. 577,567.
26. 766,475.
27. 376,737.
28. 875,565.
29. 666,567.
30. 777,657.
31. 267,146.
32. 365,662.
33. 307,177.
34. 176,515.
35. 262,607.
36. 421,256.
37. 277,465.
38. 219,173.
39. 373,676.
40. 378,074.
41. 257,156.
42. 471,465.
43. 517,774.
44. 76,516.
45. 311,856.
46. 677,172.
47. 376,046.
48. 562,567.
50- 1. 40,895.
2. 42,073.
3. 41,217.
4. 44,236.
5. 42,308.
51- 6. 44,509.
7. 41,808.
8. 46,765.
9. 46,558.
10 48,802.
52-11. 4,628.

12. 8,462.
13. 4,028.
14. 60,248.
15. 63,906.
16. 68,402.
17. 46,082.
18. 70,492.
19. 127,068.
20. 92,704.
21. 106,926.
22. 109,356.
23. 79,359.
24. 92,704.
25. 193,578.
26. 214,496.
27. 193,572.
53- 28. 213,704.
29. 258,144.
30. 71,284
31. 139,056.
32 145,704.
33. 105,738.
34. 218,712.
35. 219,276.
36. 387,156.
37. 231,760.
38. 254,096.
39. 140,168.
40. 202,680.
41. 182,772.
42. 230,175.
43. 157,824.
44. 150,216.
45. 312,384.
59- 1. 52,279.
2. 45,897.
3. 57,437.
4. 50,384.
5. 51,459.
60- 6. 47,931.
7. 54,123.
8. 54,794.
9. 57,547.
10. 56,480.
64- 1. 7,867,676.

2. 73,276,956.
3. 79,097,578.
4. 71,178,967.
5. 87,819,971.
6. 8,765,687.
7. 68,797,784.
8. 5,748,957.
9. 73,287,865.
10. 86,995,768.
11. 80,773,478.
12. 26,579,678.
65-13. 7,170,658.
14. 94,880,045.
15. 860,697,867.
16. 47,159,685.
17. 32,660,068.
18. 6,697,868.
19. 77,995,682.
20. 90,859,567.
21. 77,706,580.
22. 93,730,145.
23. 35,764,940.
24 70,299,847.
25. 77,660,078.
67- 1. 246,722.
2. 296,492.
3. 213,852.
4. 402,241.
5. 249,494.
6. 281,968.
7. 459,704.
8. 211,476.
9. 368,464.
68-10. 5,480,296.
11. 3,895,469.
12. 5 142,276.
13. 2,548,232.
14. 2,548,546.
15. 29,126,064.
16. 53,085,480.
17. 1,483,264.
18. 1,295,866.
19. 84,320.
20. 15,074,618.
21. 1,586,880.

FIRST STEPS AMONG FIGURES. 177

22. 2,298,402.
23. 24,593,902.
24. 41,803,168.
25. 4,828,052.
26. 313,225.
27. 4,252,262.
69-28. 22,517,838.
29. 3,186,721.
30. 3,430,458.
31. 11,993,364.
32. 27,078,591.
33. 47,923,624.
34. 2,011,205.
35. 16,799,022.
70- 1. 3,423.
2. 204,307.
3. 820 306.
4. 3,060,905.
5. 520,709.
6. 1,060,907.
7. 8,050,907.
8. 64,083,207.
9. 209,071,205.
10. 7,062,304.
11. 508,207.
12. 53,208,409.
71-13. 2,457 [1-3].
14. 3,543 [1-4].
15. 5,475 [2-3].
16. 274,658 [1-3].
17. 545,734 [3-4].
18. 265,435 [3-5].
19. 238,636 [3-4].
20. 1,887,677 [1-3].
21. 6,543,346 [1-4].
22. 54,764 [3-5].
23. 154,264 [4-6].
24. 04,857 [1-4].
25. 623,542 [3-6].
26. 75,246 [3-5].
27. 543,452 [4-7].
28. 974,658.
29. 042,455 [4-7].
30. 046,819 [3-6].
31. 786,538 [1-4].

32. 24,635 [3-7].
33. 3,456,827 [3-6].
76- 1. 63,237.
2. 61,205.
3. 63,924.
77- 4. 66,698.
5. 65,307.
6. 55,344.
7. 52,771.
8. 57,783.
78- 9. 61,074.
10. 60,493.
79-19. 69,274.
20. 69,883.
21. 68,111.
22. 71,276.
23. 67,740.
80- 1. 17,438,656.
2. 23,971,752.
3. 244,806,912.
4. 382,698,325.
5. 23,698,382.
6. 26,155,500.
7. 2,595,136.
8. 4,188,375.
9. 5,965,938.
10 3,235,848.
11. 55,044,555.
12. 17,508,384.
13. 26,435 [3-4].
14. 635,241 [3-5].
15. 354,624 [1-4].
16. 3,746,254 [3-6].
17. 657,342 [3-4].
18. 4,635,246 [4-5].
81-19. 7,562,435 [3-4].
20. 437,520 [4-6].
21. 4,537,264 [3-7].
22. 5,743,263 [4-6].
23. 5,702,474 [4-6].
24. 2,605 [1-4].
25. 6,370 524 [4-7].
26. 423,177.
27. 6,258,085.
28. 3,004,756 [3-7].

29. 4,053,702 [4-6].
30. 7,086,534 [3-4].
31. 5,768,430 [4-7].
32. 6,673,048 [3-6].
33. 2,540,867 [4-6].
 1. 4,352 [143].
 2. 3,624 [177].
 3. 4,235 [214].
 4. 4,236 [1,072].
82- 5. 5,342 [447].
 6. 24,353 [1,408].
 7. 645 [1,125].
 8. 3,624 [331].
 9. 3,425 [346].
 10. 4,726 [378].
 11. 3,426 [2,139].
 12. 3,564 [2,163].
 13. 6,457 [341].
 14. 54,673 [1,342].
 15. 4,756 [3,000].
 16. 57,643 [216].
 17. 3,745 [108].
 18. 4,576.
 19. 74,656 [391].
 20. 46,576 [307].
 21. 6,534 [279].
 22. 76,487 [1,279].
 23. 7,458 [6].
 24. 64,786 [301].
 25. 5,867 [1].
 26. 4,758 [124].
 27. 6,785 [961].
 28. 65,847 [346].
 29. 47,664 [2,386].
83-30. 71,414.
 31. 71,083.
 32. 72,225.
 33. 70,304.
 34. 68,796.
89- 1. 549 bu.
 2. 370 cts.
90- 3. 864 bu.
 4. 174 marbles.
 5. 632 bu.
 6. 189.

 7. $12,250.
 8. 127 marbles.
 9. 105 cts.
 10. 1,781 steps.
 11. $85.
 12. 7,682 cts.
 13. $362.
 14. 92 da.
91-15. 425 A.
 16. $668.
 17. 91 marbles.
 18. $1,920.
 19. 25,092 cts.
 20. 59 bu.
 21. 15 cts.
 22. 874 bu.
 23. 152 bu.
92-24. 1,784,910.
 1. 64,482.
 2. 68,035.
 3. 69,530.
 4. 66,964.
 5. 68,728.
93- 6. 71,207.
 7. 75,781.
 8. 74,612.
 9. 72,156.
 10. 69,201.
94-11. 3,608,157.
 12. 658,280.
 13. 616,178.
 14. 206,078.
 15. 6,050,993.
 16. 391,648.
 17. 407,708.
 18. 278,079.
 19. 277,164.
 20. 391,667.
 21. 622,678.
 22. 578,238.
 23. 39,213,566.
 24. 670,781.
 25. 3,764,877.
 26. 79,088.
 27. 24,147.

FIRST STEPS AMONG FIGURES.

28. 42,869,909.
29. 69,059,978.
30. 37,523,976.
31. 3,189,825.
32. 36,970,172.
1. 70,919,205.
2. 47,009,130.
95- 3. 5,449,025.
4. 302,421,261.
5. 31,269,744.
6. 392,262,525.
7. 3,756,608.
8. 726,525.
9. 187,044.
10. 7,221,816.
11. 5,498,265.
12. 5,448,946.
13. 4,231,810.
14. 35,677,072.
15. 64,367,973.
16. 49,849,081.
17. 610,079,611.
18. 308,940,676.
19. 28,681,464.
20. 379,544,763.
21. 35,695,072.
22. 68,097,323.
23. 58,390,182.
24. 379,239,951.
25. 6,257,507,544.
26. 6,328,767.
96-27. 2,816,029.
28. 708,709.
29. Impossible.
1. 463,057 $^{3-7}$.
2. 634,750 $^{6-8}$.
3. 475,307 $^{1-6}$.
4. 5,463,060 $^{4-9}$.
5. 564,037 $^{3-9}$.
6. 374,675 $^{4-9}$.
−7. 465,736 $^{3-8}$.
− 8. 6,537,645 $^{6-8}$.
9. 6,870,657 $^{6-7}$.
10. 4,867,690 $^{5-9}$.
11. 6,578,648 $^{3-9}$.

−12. 5,746,387 $^{4-7}$.
13. 7,468,576 $^{7-9}$.
14. 67,580,760 $^{6-9}$.
15. 4,637,586 $^{7-9}$.
16. 65,748,760 $^{3-7}$.
17. 796,850,809 $^{6-9}$.
−18. 697,879,680 $^{1-7}$.
19. 64,859,760.
20. 745,695 $^{4-7}$.
21. 479,568 $^{1-7}$.
97-22. 840,967 $^{4-8}$.
23. 639,408 $^{7-9}$.
24. 4,253 ^{970}.
25. Omit this example.
26. 35,246 ^{103}.
27. 5,264 ^{2,143}.
28. 6,423 ^{1,243}.
29. 763 ^{12,217}.
30. 4,536 ^{1,034}.
31. 64,352 ^{341}.
32. 48,372 ^{346}.
33. 57,362 ^{3,762}.
34. 41,572 ^{303}.
35. 3,754 ^{1,000}.
36. 4,827 ^{314}.
37. 36,472 ^{347}.
38. 64,727 ^{4,043}.
39. 463,752 ^{4,800}.
40. 74,635 ^{936}.
41. 4,605 ^{326}.
42. 70,534 ^{3,041}.
43. 4,653 ^{1,124}.
44. 64,075 ^{809}.
45. 736,502 ^{123}.
98-46. 475,630 ^{2,846}.
47. 4,607 ^{3,762}.
48. 60,835 ^{81}.
49. 72,506 ^{110}.
50. 43,072 ^{3,000}.
51. 4,073 ^{10,301}.
52. 47,260 ^{470}.
53. 58,240 ^{740}.
54. 47,050 ^{79}.
55. 70,648 ^{2,876}.
56. 4,750 ^{177}.

57. 697 [247].
58. 758 [447].
59. 579 [422].
60. 796 [527].
61. 975 [241].
62. 768 [123].
63. 687 [541].
64. 876 [200].
65. 8.607 [301].
99–66. 6.908 [745].
67. 8.097 [476].
68. 7.906 [473].
69. 47,650 [330].
70. 76,305 [201].
71. 67,057 [2-245].
72. 57,642 [767].
102- 1. $1,483.
2. 726.525.
3. 7,592.
4. 7 bbl.; 4 gal. left.
5. 684 gal.
6. $6,845.
7. 197 mi.
8. $661.
9. 7.126.
103-10. 426.
11. $1,050.
12. 109 sheep.
13. Lost $900.
14. 1,796,256 apples.
15. $4,543.
16. 59,888,684.
17. 136 tons.
18. 337,022.
104-19. $19.
20. 191,284 yd.
21. Lost $50. [rem.
22. 14 horses and $50
23. $384.
24. 421 A.; $37,309.
25. $6 [54-135].
26. $4,456.
105-27. 3,743,520 ft.
28. 33 mi.
29. Nothing.

30. 86 yrs.
31. 840,413,978.
106- 1. 79,984.
2. 76,908.
3. 72,016.
4. 87,047.
5. 89,229.
107- 6. 74,077.
7. 81,212.
8. 75,464.
9. 81,195.
10. 75,825.
108-11. 79,455.
12. 82,270.
13. 84,586.
14. 78,351.
15. 78,548.
109-16. 95,050.
17. 97,295.
18. 97,187.
19. 98,616.
20. 97,489.
110- 1. 4,246,909.
2. 8,737,469.
3. 389,385,828.
4. 73,165,668.
5. 29,730,808.
6. 7,080,807.
7. 291,864,906.
8. 77,927,921.
9. 730,868,068.
10. 20,679,828.
11. 7,779,807.
12. Impossible.
13. 39,299,876.
14. 421,769,927.
15. 670,796,919.
16. 6,479,979,692.
17. 4,585,944.
18. 5,955,446.
19. 1,540,080.
20. 772,926,042.
21. 379,239,951.
22. 550,842,699.
23. 6,953,140,080.

FIRST STEPS AMONG FIGURES. 181

24. 490,619,928.
25. 4,311,281,464.
111-26. 78,650,000.
27. 5,730.
28. 6,800.
29. 6,320,000.
30. 87,500,000.
112-31. 172,500,000.
32. 16,205,000.
33. 444,500,000.
34. 18,700,000,000.
35. 4,336,400,000.
36. 17,983,000.
37. 51,300,000.
38. 6,887,200.
39. 59,073,000. ;
40. 40,843,000.
41. 3,407,454,000.
42. 452,276,640,000.
43. 3,589,638,200,000.
44. 4,827,581,000.
45. 6,409,800,000.
  1. 4,798,697 [5-9].
  2. 87,806,790 [7-9].
113- 3. 769,589,079 [5-6].
  4. 9,687,890 [2-7].
  5. 769,809,780 [8-9].
  6. 6,589,679 [6-8].
  7. 978,697,089 [6-9].
  8. 76,898,697 [4-6].
  9. 7,049,680 [6-7].
  10. 4,906,704 [7-9].
  11. 79,684,796 [6-7].
  12. 958,007,980 [5-8].
  13. 893,798,400 [7-8].
  14. 65,870,486 [6-7].
  15. 870,956 [8-9].
  16. 75,680,390 [6-7].
  17. 903,780 [8-9].
  18. 9,586,090 [7-9].
  19. 89,607,980 [8-9].
  20. 5,869,759 [6-7].
  21. 4,375 [6,901].
  22. 5,786 [821].
  23. 365 [310].

24. 147,283 [7,000].
25. 6,485 [700].
26. 46,372 [625].
114-27. 14,735 [449].
28. 17,399 [321].
29. 68,591 [7,457].
30. 147,964 [946].
31. 9,608 [4,879].
32. 13,568 [274].
33. 49,807 [291].
34. 138,769 [3,700].
1. 79 [2,466].
2. 876 [80].
3. 23 [47,600].
4. 762,196 [46].
5. 37 [1,936,641].
116- 1. 486 [3,211].
2. 6,857 [12,131].
3. 796 [76,900].
4. 96 [21,362].
5. 873 [4,756].
6. 98 [90,000].
7. 468 [2,300].
8. 2,005,000,000.
9. 7,958 [122].
10. 68,050 [750].
11. 7,960 [10].
12. 5,920,005,998.
13. 48,690 [20,172].
14. 79,080 [700].
117-15. 1,785 [76,498].
16. 60,970 [36].
17. 9,780 [56].
18. Impossible.
19. 740 [187,000].
20. 8,069 [37,231].
21. 6,095 [300].
22. 405,025,000.
23. 78,096 [231].
24. 796,000.
  1. 947,004.
  2. 767,679.
  3. 9,058,032.
  4. 10,149,216.
  5. 749,042,756.

KEY TO BEEBE'S

6. 7,043,688.
118- 7. 78,956,244.
8. 833,855,616.
9. 104,154,369.
10. 5,755,896.
11. 1,171,075,764.
12. 83,710,764.
13. 837,443,940.
14. 1,136 156,268.
15. 11,758,752.
16. 7,543,756.
17. 105,587,748.
18. 598,772,364.
19. 1,175,855,832.
20. 95 879,820.
21. 98,790,948 [1-12].
22. 49,688,132 [9-11].
23. 68,374,969 [6-12].
24. 87,076,859 [7-12].
25. 786,547,997 [8-12].
26. 869,897,046 [9-12].
27. 37,998,060 [10-12].
28. 79,684,968 [9-12].
29. 74,869,740 [11-12].
30. 675,846,090 [2-11].
125- 1. $2,010.
2. $34,995.
3. 490,000.
126- 4. 17 yrs.
5. 236 A.; $74 [136, 226].
6. 15,148.
7. $42.
8. 123,685.
9. 38 mi.
10. 60 cts.
11. 3,552; 3,066.
12. $760.
127-13. $14,352. [& $27 left.
14. 174 [27-57] A.; or 174 A.
15. 614 mi.
16. 22,999,800,925.
17. $2,250.
18. 22,032 solid ft.
19. $4,340.
128-20. $8,380.

21. $80.
22. $80.
23. $945.
24. 192 da.
25. $322.
26. 594 ml.; 54 ml.
27. 136 bu.
28. $45.
129-29. 118 [79-136]. [rem.
30. 60 horses, and $53
31. 42 times.
32. 183 bu.
33. $82.
34. 72 weeks.
35. 581 sheep.
36. 429 cts.
37. 398 lbs.
130-38. 15 sheep.
39. 49 animals.
40. 6 yrs.
41. 1,440 lbs.
42. 3,536.
43. $772. [rem.
44. 6 horses and $40
45. 1,274 bags.
131-46. 32 teachers & $175 remaining.
47. 14 cows.
48. 949 cts.
49. Gained 80 cts.
50. 70 half dimes.
51. 37 bags; 518 cts.
52. 132 birds.
132-53. 1,162,568 [800-1,000] sec.
54. $4,730.
55. 48.
Omit "more."
56. Lost $685.
57. $86.
58. 6,784. [left.
59. 76 Co's and 15 men
60. 27 yds.
133-61. 1,215 rd.; 9,651 rd.
62. 12 yrs.
63. 158.

64. 8,589,783. [902.
65. Quo. 105; rem. 3,-
66. 2,392.
135- 1. 70,169.
2. 67,005.
3. 58,369.
4. 74,595.
5. 74,473.
136- 6. 76,967.
7. 81,025.
8. 73,892.
9. 72,772.

10. 74,617.
137-11. 90,190.
12. 95,621.
13. 93,522.
14. 89,651.
15. 93,822.
138-16. 104,643.
17. 108,075.
18. 98,259.
139-19. 111,157.
20. 111,236.

APPENDIX.

DETAILED METHODS IN ARITHMETIC.

FROM THE COURSE OF STUDY PREPARED FOR THE PUBLIC SCHOOLS OF SAN FRANCISCO.

I. LESSONS FOR BEGINNERS.

Grube's Method in Number.

THE following are in substance some of the most important principles given by Grube for his method in teaching beginners to comprehend numbers and their relations.

Principles.

" 1. Each lesson in Arithmetic must also be a lesson in language. The teacher must insist on readiness and correctness of expression. As long as the language for the number is imperfect, the idea of the number will be defective.

" 2. The teacher must require the scholar to speak as much as possible.

" 3. Answers should be given occasionally

by the class in concert, but usually by the scholar, individually.

"4. Every process must be illustrated by means of objects.

"5. Measure each new number with the preceding ones.

"6. Teachers must insist on neatness in making figures."

ORDER OF STEPS.

First Step. Illustrate the required combinations by means of *counters*, such as blocks, splints, or shells, in the hands of the children themselves, and by other objects in the hands of the teacher.

Second Step. Express the same combinations on the blackboard or on slates with *marks*.

Third Step. Take the same combinations mentally with abstract numbers.

Fourth Step. Practical problems in applied numbers.

HOW TO BEGIN.

☞ The time required for the work will depend upon the age of the children, as also somewhat upon their natural ability. Some children may require a year to complete the work which others may master in a term.

I. *THE NUMBER ONE.*

1. Hold up one counter, one hand, one finger, one slate, etc.

On your slate make a straight mark, one dot, one cross, etc.

On the blackboards make one mark, one dot, one cross, etc.

2. Place one counter in the middle of the desk ; take it away ; how many have you left ?

Make one mark on your slate ; rub it out ; how many marks are left ?

3. Send the class to the blackboards and let them make the mark for one thus, | ; and also the figure thus, 1.

4. Proceed very slowly. Much time should be given to those who do not learn easily.

II. *THE NUMBER TWO.*

1. Each of you take one counter and place it by itself on your desk ; now take another,

and place close to it; how many counters have you? (Require the answer in a full sentence.)

Make one straight mark on your slate; make another close to it; how many have you now?

Go to the blackboard: make one mark; another, close to it; how many now?

Clap your hands once; again; how many claps?

Rap on your desk once; again; how many raps?

2. *Counting.*—Place one counter on your desk, *; a little way off from the first one, place two counters close together, thus * *. Count, *one two ; two, one.*

On your slates make marks thus, | || , and count forwards and backwards.

3. *Addition.*—I. Place one counter on the desk; place another counter close to it; how many have you now? *Ans. I have two counters.* How many counters are one counter and one counter? . *Ans. One counter and one counter are two counters.* [The teacher will further illustrate with books, pencils, crayons, etc.]

II. *Slate and Blackboard.*—Make one mark; another one near it. How many marks have you made?

[Continue with rings, dots, crosses, etc.]

4. *Subtraction.*—I. Place two counters together on your desk; take one away; how many have you left? *Ans. I have one left.* One counter from two counters leaves how many? *Ans. One counter from two counters leaves one counter.*

[Teachers will continue with fingers, hands, books, and other objects.]

II. *Slate and Blackboard.*— Make two marks; rub out one; how many are left? Make two marks; rub them out; how many are left? *Ans. None are left.* Two taken away from two leaves how many?

5. *Multiplication.*—I. Each of you put one counter on the desk; now put another one with it; how many times have you taken one counter? *Ans. I have taken one counter twice.* Two times one counter are how many counters? *Ans. Twice one counter are two counters.*

II. *Slate and Blackboard.*—Make one mark; now another. How many times have you made one mark? *Ans. I have made one mark twice.* Then two times one mark are how many marks? *Ans. Two times one mark are two marks.*

6. *Division.*—See Teacher's Edition of First Steps, page 14, paragraph 10.

7. *Comparison.*—Give one counter to John and two to Frank. How many counters has John? Frank? How many has Frank more than John? How many more is two than one?

How many counters has John less than Frank? Then one is one less than two, and two less than two is nothing.

Blackboards. — Illustrate the same with marks.

General Remarks.

It is a feature of this method, that it teaches by the eye as well as by the ear, while in most other methods arithmetic is taught by the ear alone. If a child is to measure 7 by the number 3, the illustration, by comparison is :

```
 *   *   *
 *   *   *
 *
```

".If counters are arranged in this way, and impressed upon the child's memory as depicting the relation between the number 3 and 7, it is, in fact, all there is to know about it. Instead of teaching all the variety of possible combinations between 3 and 7, it is sufficient to make the child keep in mind the above picture. The first four rules, as far as 3 and 7 are concerned, are contained in it, and will result from expressing the same thing in different words, or describing the picture in different ways. Looking at the picture, the child can describe it as :

$3+3+1=7$, or $3 \times 2+1=7$, or $7-3-3=1$, $7 \div 3 = 2$ (1). The latter process is to be read : 3 in 7 twice, and 1 remaining.

"Let the number to be measured be 10, and the number by which it is to be measured be 4 ; then the way to arrange the dots is :

```
*   *   *   *
*   *   *   *
*   *
```

"The child will be able to see at once, by reading, as it were, that $4+4+2=10$, $4 \times 2 +2=10$, $10-4-4=2$, $10 \div 4 = 2$ (2), and to perceive at a glance a variety of other combinations. The children will, in the course of time, learn how to draw these pictures on their slates in the proper way. Nor will it take long to make them understand that every picture of this kind is to be 'read' in four ways, first using the word *and*, then *times*, then *less*, then *in.* As soon as the pupils do this, they have mastered the method, and can work independently all the problems, within the given number, which are required in measuring."

Order of Steps.

I. Counters.
II. Figures.
III. Abstract Numbers.
IV. Practical Problems.

FIRST STEPS
AMONG FIGURES.

A Drill Book in the Fundamental Rules of Arithmetic.

Pupils' Edition.

BY

LEVI N. BEEBE,

CANANDAIGUA, N. Y.

SYRACUSE, N. Y.:
C. W. BARDEEN, PUBLISHER.
1881.

COPYRIGHT, 1877, LEVI N. BEEBE.

PREFACE TO REVISED EDITION.

Great care has been taken to correct the errors of the first edition and it is hoped that few remain.

At the request of several teachers in ungraded schools, the addition, subtraction, multiplication, and division tables have been inserted in the Pupils' Edition, and a number of pages have been prefixed to relieve teachers in such schools of much of the labor of oral instruction. In graded schools the pupils should not have a book until they have been taught orally as far as to the tables of 7's and review, p. 29.

The Teachers' Edition contains the answers to the examples in this book, and instruction for oral work with the youngest pupils, together with methods and additional examples. It is, therefore, necessary that the teacher should have it and carry along the work of the two editions together. References at the bottom of the pages in each edition call attention to the pages of the other edition which contain work of the same kind.

The Pupils' Edition is bound separately for

pupils' use, as well as with the Teachers' Edition for teachers' use.

The object of having the double book, the Teachers' Edition and the Pupils Edition, is that while the pupil is to prepare his lesson from the Pupils' Edition, the teacher has, in the Teachers' Edition, additional examples which the pupil has not seen, which are intended to be assigned for solution during recitation as a test of his knowledge of the subject in hand.

The first 96 pages of the Teachers' Edition, together with the parallel work of the Pupils' Edition to p. 44, are bound separately for the use of teachers in the oral instruction of pupils during their first two or three years in graded schools. It is intended wholly for oral work and is called "First Steps Among Figures, Oral Edition."

Much care has been taken in each edition to proceed from the easiest examples to those that are more difficult, in order to avoid discouraging the pupil.

The author is indebted to his assistant teachers for aid in the preparation of the very large number of examples in the book.

For a more extended notice of the scope and plan of this work see the preface to the Teachers' Edition, and also the Special Notice which precedes it.

LEVI N. BEEBE.

CANANDAIGUA, N. Y., *April*, 1878.

SPECIAL NOTICE.

Persons who may examine this book are asked to notice especially the following: Examples for rapid solving on pp. 17, 20, 25, 29, 38, 41, 45, 51, etc., as well as the foot notes, pp. 16 and 17; the drill in reading and writing numbers on pp. 39, 47, 53, 60, 69, 83, 111 to 116, etc.; the series of division with remainders, which is to prepare the pupil for short division, pp. 107, 128, 129, 145, and 157. In the examples in long division, on p. 136, Teachers' Edition, and on p. 82 of the Pupils' Edition, since the second figure from the left in the divisor is a cipher, while the figures of the quotient are small, the divisor is contained in each partial dividend just as many times as it appears to be. For instance, the first divisor in the Pupils' Edition, p. 81, is 201; and if the reader doubts that the path is thus made easy for the beginner, let him give his pupils an example with 19, 29, or 291 for a divisor, and then one with 201 as a divisor. There are no ciphers in the quotients on those pages, so that every difficulty is postponed to a later time that can be so put off. The examples were made by assuming such a divisor and quotient as were desirable, multiplying them, and adding an assumed remainder to the product, which gave the dividend found in the book.

After the practice on the twelve examples on p. 136 in the Teachers' Edition, and on the twenty-nine examples on p. 82 of the Pupils' Edition, there are a number of pages of other work, after which long division, with the same sort of easy examples, recurs on p. 146, Teachers' Edition, and p. 97, Pupils' Edition. After a few examples the cipher occurs in the quotient, of which the teacher is warned at the bottom of p. 146, Teachers' Edition. Additional difficulties are treated on pp. 147 and 148, followed by examples in illustration, and still others on pp. 114, 115, and 116, Pupils' Edition.

The continuous form commonly used for tables of addition, subtraction, multiplication, and division, (as 2 and 2 are 4, 3 and 2 are 5, 4 and 2 are 6, etc.), has been forsaken for the form found on pp. 33, 37, 43, 48, 56. 63, etc., Teachers' Edition, and on pp. 6, 9, 18, 24, 29, etc., Pupils' Edition.

If the teacher copies the series on the blackboard, he may write the answers underneath, or require the pupils to find the answers, as he prefers.

For instructions see pp. 42 and 43, Teachers' Edition, and pp. 5, 6, and 19, Pupils' Edition.

The teacher will see that at p. 63 he should begin to use also the Pupils' Edition, even though the pupil has not yet obtained his book. The number at the bottom of that page refers to the page of the Pupils' Edition that has the same kind of work. The numbers at the bottom of the following pages in the Teachers' Edition refer in the same way, while those in the Pupils' Edition refer back to the Teachers' Edition. The work of the two editions after reaching these parallel pages should be carried along together carefully.

The operations of addition, subtraction, multiplication, and division being taught together throughout the whole book, beginning with the easiest examples and progressing gradually to more difficult ones, work in each rule is constantly recurring. This necessitates a peculiar arrangement of the work, but it constitutes one of its chief excellencies.

The Pupils' Edition is bound separately for pupils' use.

The key containing answers will be found on pp. 171-183, Teachers' Edition.

For a still further description of the plan of the book please read the preface of each edition

ADDITION TABLE. *5's and review*.

The following table is best learned by repeating each set of numbers many times, as: "2 and 4 are 6," "2 and 4 are 6," etc., until it is well learned. "*Oft repeated, long remembered.*" In reciting, it is best not only to say 2 and 4 are 6, but also 2 from 6 leaves 4, and so on through the table, or at least until the pupil sees clearly that subtraction is the opposite of addition. If the table is copied on the blackboard without the answers, the recitation may be conducted as in oral spelling, each pupil reciting a section, and any error being corrected in his turn by the first pupil who has noticed it. The teacher may ask the questions directly from the table in the book, if he prefers it.

The letters are names for the sections, and are a convenience in assigning the lesson.

When the pupils have learned the following table, test their knowledge of it by the table in Teachers' Edition, p. 63, which is differently arranged.

a			*b*			*c*		
4	2	5	3	1	4	2	5	3
2	3	4	5	2	3	4	5	2
6	5	9	8	3	7	6	10	5

d			*e*			*f*			*g*	
2	4	1	5	3	1	4	2	5	3	1
5	4	3	2	3	4	5	2	3	4	5
7	8	4	7	6	5	9	4	8	7	6

"2 and 4 are 6" may be written 2+4=6, and is read "2 plus 4 equals 6." (Always recite from below upward, that is 2 and 4, not 4 and 2).

1. George has 3 cents and Mary has 5 cents; how many have both?

Solution: They have the sum of 3 cents and 5 cents, or 8 cents.

2. John has 4 marbles and Henry has 2; how many have both?

3. Susan had 5 pins and afterward found 4; how many had she then?

4. A good boy brought in 4 armfuls of wood for his mother in the morning, and 3 armfuls in the afternoon; how many armfuls did he bring in that day?

5. Nettie found 2 eggs in one nest and 5 in another; how many did she find in both?

6. Helen had 3 sleigh-rides on Monday and 3 on Tuesday; how many did she have in the two days?

7. Carrie had 1 needle and her mother gave her 5; how many had she then?

8. Charles rode down hill on his own sled 4 times and on his brother's sled 4 times; how many times did he ride down hill?

9. Count by 2's from 2 to 10, thus: 2, 4, 6, 8, 10.

10. Read the following numbers, or write them in words and bring them to the recitation: 76; 94; 58; 85; 67; 93; 72; 25; 61; 38; 15; 88; 29; 13; 36; 11; 21; 12; 34; 98; 14.

11. Count by 2's from 2 to 20.

Write in figures (Arabic): (12.) forty-six; (13.) fifty-three; (14.) eighty-one; (15.) sixty-nine; (16.) thirty-five; (17.) twenty-seven; (18.) fifty; (19.) seventeen; (20.) twelve. The teacher should give more exercise of this kind.

21. What cost an orange at 5 cents and a lemon at 3 cents?

22. Walter bought one book for 2 shil-

lings and another for 4 shillings; how much did he pay for both?

23. An inattentive pupil whispered 3 times in the forenoon and 4 times in the afternoon; how many times did he whisper during the day?

24. On Nellie's birthday her father and mother each gave her five shillings; how much money did both give her?

25. Count by 2's from 2 to 60.

26. Write in figures (Arabic): VII; II; IV; I; VIII; VI; IX.

27. Write in letters (Roman): 5; 9; 2; 4; 7; 10; 3; 6; 1; 8.

28. Count by 2's from 1 to 9.

29. Count by 2's from 1 to 13.

30. Add 2, 2, 1, 2, 2, 1, 2, 2. The teacher will teach the pupil to write the numbers on his slate in a column, with a line underneath, and to write the answer beneath the line.

31. Add 1, 2, 2, 2, 1, 2, 1, 1, 2, 2.

32. Add 2, 2, 2, 1, 2, 2, 1, 2.

33. Add 1, 1, 1, 2, 2, 2, 2, 2.

SUBTRACTION TABLE.* *5's and review.*

"*Oft repeated, long remembered.*"

a			b			c		
6	7	4	8	7	5	10	5	9
3	2	3	4	5	2	5	4	5
3	5	1	4	2	3	5	1	4

d			e			f			g	
4	8	7	6	5	6	9	8	3	7	6
2	3	4	5	3	2	4	5	2	3	4
2	5	3	1	2	4	5	3	1	4	2

"3 from 6 leaves 3" may be written 6—3=3, and is read "6 minus 3 equals 3."

After this table is learned, test the class by using the table in Teachers' Edition, p. 63, which is differently arranged.

1. Count by 2's from 1 to 19.

2. Jesse having 7 cents spent 3 of them; how many had he left?

Solution: He had left the difference between 7 cents and 3 cents, or 4 cents.

* The teacher will treat this table much as he did the previous addition table. See the instructions with that table.

3. Edward had 9 cents; he lost four of them; how many had he left?

4. Albert had 5 marbles, and gave John 2 of them; how many had he left?

5. William has 10 cents and Lewis has 5 cents; how many more has William than Lewis?

6. Eight boys were skating on the ice, when three of them fell; how many remained standing?

7. Count by 2's from 1 to 29.

8. Read the following numbers: 100; 163; 175; 158; 189; 107; 309; 246; 350; 416; 112; 761; 468; 906; 413; 970; 320; 518; 971; 800.

Write the following numbers in figures (Arabic): (9.) four hundred sixty-three; (10.) two hundred thirty-nine; (11.) seven hundred eighty-one; (12.) one hundred sixteen; (13.) five hundred eight; (14.) six hundred ninety; (15.) seven hundred twelve; (16.) three hundred twenty-one; (17.) five hundred; (18.) nine hundred two; (19.) five hundred sixty-one; (20.) one hundred seven; (21.) the number of days in a year; (22.) eight hundred; (23.) one hundred nineteen; (24.) nine hundred forty.

25. Count by 2's from 1 to 69.

See Teachers' Edition, p. 40.

Write in figures (Arabic): (26.) XV;
(27.) XXII; (28.) XIX; (29.) XXIX;
(30.) XXIII; (31.) XVI; (32.) IX; (33.)
XXIV; (34.) XIV.

Write in letters (Roman): (35.) twelve;
(36.) nineteen; (37.) fifteen; (38.) twenty-four; (39.) twenty-two; (40.) twenty-five; (41.) twenty-three.

42. Count by 3's from 3 to 15.
43. Add 1, 2, 2, 2, 2, 2, 1, 2, 2, 2.
44. Add 2, 1, 2, 1, 2, 2, 1, 2, 2, 2.
45. Add 2, 2, 2, 1, 2, 2, 2, 2, 2, 2.
46. Add 2, 2, 1, 2, 2, 1, 1, 2, 2, 2.

Make and solve several examples like the preceding ones.

47. A dwarf tree had 7 pears on it, but 4 of them fell off; how many remained on the tree?

48. Samuel's mother gave him 8 cents to spend; he paid 4 cents for candy, and bought marbles with the remainder; how much did he spend for marbles?

49. Carlos started for school with 9 marbles in his pocket, but when he got there he found he had lost all but 4 of them; how many had he lost?

50. Maggie had 6 needles; she broke 3 of them; how many were unbroken?

51. How many more wheels has a carriage than a sulky?

52. How many less fingers on one hand than on both?

53. 8 is how many more than 3?

54. Mr. Jones was idle 3 days of a week; how many days of the week did he work?

55. Mr. Rawson worked 5 days one week and 4 the next; how many days did he work in the two weeks?

56. Julia misspelled 2 words in the forenoon and 4 in the afternoon; how many did she misspell that day?

57. Joseph's mother told him he might eat 2 apples; he ate 5 apples; how many more did he eat than he ought?

58. From Mr. Collins's house to the post-office it is 4 miles, and from the post-office to the school-house it is 3 miles farther; how far is it from Mr. Collins's house to the school-house?

59. Count by 3's from 3 to 24.

60. Add 1, 2, 2, 2, 2, 3, 3, 2, 2, 2.

61. Add 2, 2, 2, 1, 2, 2, 2, 2, 2, 2, 2, 2, 2, 2.

62. Add 1, 2, 2, 1, 1, 2, 2, 1, 2, 2, 2, 1, 2, 2.

63. Add 1, 2, 2, 2, 1, 2, 2, 2, 1, 1, 1, 2, 2, 2.

64. Add 2, 2, 2, 2, 1, 2, 2, 2, 2, 2, 1, 1, 1, 2, 2.

See Teachers' Edition, p. 66.

65. Add 1, 2, 1, 2, 3, 3, 3, 1, 2, 3, 3, 2, 1, 2, 2.

66. Add 2, 2, 1, 2, 2, 2, 2, 2, 3, 3, 2, 2, 2.

67. Susan took 3 peaches from a pile of 6 peaches; how many were left in the pile?

68. A boy was sent to the store with 8 cents; he spent all of them but 3; how many did he spend?

69. William spent 2 cents for nuts and 5 cents for licorice; how much did he spend altogether?

70. John hit James 4 times with snowballs and James hit John twice; how many more times was James hit than John?

71. In a street car are 4 women and 5 men; how many persons in the car?

72. There were 2 pictures on one side of a wall and 3 on the other; how many on both?

73. James had 8 doves; 2 pairs of them were killed; how many lived?

74. A little girl has a cushion with 6 pins on it; she took off 3 of them; how many were left?

75. A little boy earned 2 cents; he then had 7 cents; how many had he at first?

76. In a geography class, 4 pupils were in order and 3 were not in order; how many pupils in the class?

When a number to be read contains more than three figures, place a comma before the third figure from the last; thus, 3,216. It is read, three thousand two hundred sixteen.

Read the following numbers: (1.) 5281; (2.) 6157; (3.) 9640; (4.) 7500; (5.) 8609; (6.) 2050; (7.) 7008; (8.) 1000; (9.) 7318; (10.) 5071; (11.) 4019; (12.) 2800; (13) 5000; (14.) 8060; (15.) 7801; (16.) 1007.

The teacher should show the pupil that in writing a number that contains thousands a comma should be placed after the number of thousands, and then the remainder of the number should be written at the right, any of the three places which are not occupied being filled with ciphers.

Write in Arabic: (17.) three thousand six hundred forty-five; (18.) seven thousand nine hundred fifty-one; (19.) one thousand two hundred seventy; (20.) eight thousand three hundred; (21.) two thousand twenty; (22.) six thousand one hundred five; (23.) nine thousand; (24.) five thousand seven; (25.) one thousand

See Teachers' Edition, p. 53.

four hundred; (26.) seven thousand nine hundred four; (27.) five thousand forty; (28.) eight thousand six; (29.) four thousand.

30. Count by 3's from 3 to 60.

Write in Arabic: (31.) XXXVI; (32.) XXIV; (33.) LV; (34.) XL; (35.) LXII; (36.) LXXXVII; (37.) LXIX; (38.) LXXVI; (39.) XIX.

Write in Roman: (40.) twenty-eight; (41.) fifty-four; (42.) forty-eight; (43.) seventy-seven; (44.) eighteen; (45.) fourteen; (46.) eighty-six; (47.) thirty-two; (48.) sixty-nine; (49.) fifty-seven; (50.) sixteen; (51.) thirty-four; (52.) seventy-five; (53.) forty-seven; (54.) eighty-eight; (55.) sixty-nine.

56. Count by 3's from 1 to 61.
57. Count by 2's from 2 to 60.
58. Count by 2's from 1 to 61.
59. Count by 3's from 3 to 60.

The teacher should give a great deal of practice in counting as in the four examples above.

The teacher will show the pupil that the numbers in the following examples are to be so written in columns that the right hand figures shall be in one column. Add the right hand column first. Write below it the right hand figure of the result and add the left hand figure to the next column.

60. Add 31, 23, 30, 2, 33, 13, 22, and 2.
61. Add 22, 31, 12, 23, 3, 20, 32, 3, and 23.
62. Add 33, 20, 2, 13, 33, 30, 22, 23, and 30.
63. Add 23, 1, 33, 20, 21, 2, 10, 23, 12, and 33.
64. Add 13, 20, 23, 2, 20, 12, 3, 30, 23, and 1.
65. Add 33, 30, 2, 23, 20, 33, 3, 23, 10, 3, 32, and 12.
66. Add 32, 13, 33, 20, 12, 32, 22, 33, 21, 13, 32, and 33.
67. Add 30, 2, 33, 21, 32, 23, 3, 30, 33, 21, 3, and 2.
68. Add 23, 33, 20, 12, 3, 32, 30, 1, 13, 32, 33, and 23.

1. In a certain street there were 10 houses, but last night 5 of them were burned; how many remain?
2. Fannie had 5 pinks; she gave three of them to Jane; how many did she keep?
3. A boy could not tell how many 5 and 3 and 2 are. Can you? Show that you can.
4. Kate had a dime; she spent 10 cents for a doll; how many cents had she left?
5. There were 7 boys on a bench, 5 of

whom were studying; how many were idle?

6. Fred had 5 cents left after spending 4 cents; how many had he at first?

7. Four boys and three girls brought their lunch to school; how many do you think stayed at noon?

8. Sherman had 3 books given him on Christmas; he had 3 before; how many has he now?

9. Nine boys were playing soldier; 5 of them were called home; how many were left to play?

10. 8 apples are 4 more than James has; how many has he?

11. Harry jumped 3 feet and Jimmy jumped 2 feet farther; how far did Jimmy jump?

12. A bad boy threw 5 kittens into the pond; 3 of them swam to the shore; how many were drowned?

13. Six boys were snow-balling; two of them were hurt so they would not play any more; how many finished the game?

14. Bennie is 5 years old and Carrie is 4 years older; how old is she?

15. Add 31, 23, 21, 3, 33, 20, 12, 32, 23, 3, and 33.

16. Add 12, 3, 21, 33, 30, 23, 12, 22, 20, 32, and 3.

See Teachers' Edition, p. 67.

17. Add 3, 32, 13, 33, 20, 12, 2, 13, 22, 31, 23, and 13.
18. Add 31, 23, 33, 2, 13, 22, 30, 33, 23, 31, 22, and 33.
19. Add 2, 23, 33, 20, 13, 21, 3, 33, 20, 31, 13, 23, and 21.
20. Add 32, 13, 22, 33, 3, 21, 30, 33, 13, 22, 33, and 32.
21. Count by 3's from 2 to 62.
22. Count by 4's from 4 to 20.

MULTIPLICATION TABLE. *5's and review.*

"*Oft repeated, long remembered.*"

```
      a              b              c
   4  2  5        3  1  4        2  5  3
   2  3  4        5  2  3        4  5  2
   ─────────      ─────────      ─────────
   8  6  20       15 2  12       8  25 6

         d                    e
      2  4  1              5  3  1
      5  4  3              2  3  4
      ─────────            ─────────
      10 16 3              10 9  4

            f                    g
         4  2  5              3  1
         5  2  3              4  5
         ─────────            ─────────
         20 4  15             12 5
```

"2 times 4 are 8" may be written 4×2=8, and is read "4 multiplied by 2 is 8," or "2 times 4 equals 8."

At least some of the recitation of the above table should be as follows: "2 times 4 are 8, 2 in 8 four times;" "3 times 2 are 6, 3 in 6 twice," etc. Always recite these tables upward, that is, 2 times 4, 3 times 2, 4 times 5, etc.

After learning this table test the pupils by using the table in Teachers' Edition, p. 63, which is differently arranged.

1. John bought 5 pencils at 4 cents each; what did they cost?

Solution: If one pencil costs 4 cents, *five* pencils will cost 5 times 4 cents, or 20 cents. If preferred, take the following: They will cost 5 times 4 cents, or 20 cents.

2. What cost 3 oranges at five cents each?

3. If one sheet of paper costs 2 cents, what will 4 sheets cost?

4. A boy bought 3 marbles worth 2 cents each; what should he pay for them?

5. How many pints in 5 quarts?

Show that there are 2 pints in a quart by pouring 2 pints of water, one at a time, into a quart cup. Show in a similar manner that there are four quarts in a gallon.

See Teachers' Edition, p. 71.

Solution: In one quart there are 2 pints, in *five* quarts there are 5 times 2 pints, or 10 pints.

6. How many quarts in 3 gallons?

7. A lady bought 3 quarts of milk; she has only pint tickets; how many tickets should she give for the milk?

8. If a pig should be fed 3 ears of corn at one feeding, how many ought he to have at 5 feedings?

9. Harry has 5 cents for picking one bushel of hops; how many cents will he have for picking 3 bushels?

10. A boy's mother gave him 3 cents for each armful of wood he brought in; he brought in 4; how many cents did he earn?

11. Three pupils have each 3 books; how many books have they altogether?

12. Jane has two dolls; Mary has 3 times as many; how many has Mary?

Write in Arabic: (13.) CXI; (14.) XC; (15.) CL; (16.) CXXV; (17.) XCVI; (18.) CXIX; (19.) CLXXIV; (20.) LXXXVII; (21.) CLV; (22.) XLVIII; (23.) CI; (24.) CLXI.

Write in Roman: (25.) ninety-six; (26.) one hundred seventy-six; (27.) one hundred eight; (28.) ninety-nine; (29.) one hundred forty-eight; (30.) one hundred eighty-two.

Read the following numbers: (31.) 25758; (32.) 74051; (33.) 16300; (34.) 81407; (35.) 40520; (36.) 60000; (37.) 70030; (38.) 9010; (39.) 18002; (40.) 27643.

Write in Arabic:

41. Fifty thousand two hundred nineteen.
42. Twenty-seven thousand thirty.
43. One thousand one hundred.
44. Sixty-five thousand four hundred ninety.
45. Seventy thousand twenty.
46. Forty-six thousand ninety-one.
47. Nine thousand eight.

For more practice see Teachers' Edition, page 69.

48. Count by 4's from 4 to 60.
49. Count by 4's from 1 to 17.
50. Add 23, 33, 20, 3, 22, 13, 2, 33, 21, and 23.
51. Add 2, 33, 23, 31, 12, 3, 30, 23, 12, 33, and 32.
52. Add 33, 21, 13, 23, 3, 32, 23, 1, 23, 30, 23, and 33.
53. Add 23, 33, 30, 22, 13, 23, 30, 13, 2, 32, 33, and 20.
54. Add 3, 31, 23, 2, 33, 30, 23, 32, 22, 32, 13, and 33.

See Teachers' Edition, pp. 60, 61.

55. Add 32, 21, 33, 3, 32, 20, 31, 23, 13, 22, 31, and 2.

56. Add 2, 33, 21, 32, 13, 3, 23, 12, 32, 33, 23, 30, and 3.

57. Add 31, 12, 23, 33, 2, 23, 31, 33, 23, 22, 31, and 33.

1. In a school-room there are 5 desks in each row, and 4 rows; how many desks in the room?

2. Lester wasted 5 minutes in school every day; how much time did he waste in a week?

3. Three families of mice live in the garret; there are five mice in each family; how many mice in the garret?

4. How many quarts in 5 gallons?

5. If John picks 5 quarts of cherries one day and 4 quarts the next, how many does he pick in the two days?

6. A mother who had 5 hungry boys made 2 loaves of bread one day, 3 the next, and 4 the third day; how many loaves did she make in the 3 days?

7. There are 3 boats on the lake, and 4 boys in each boat; how many boys on the lake?

8. What cost 3 balls, if one ball cost 3 shillings?

9. James had 7 buttons on his coat; he lost off 3; how many remained on?

FIRST STEPS AMONG FIGURES. 23

10. There were 8 panes of glass in a window; a boy broke 2 of them with his ball; how many whole ones were there then?

11. A man paid 5 dollars for 2 dogs; one of them cost 3 dollars; what did the other cost?

12. Katie bought 3 yards of ribbon at 2 cents a yard to trim her doll's bonnet; how much did the trimming cost?

13. If one top costs 4 cents what will 5 tops cost?

14. I exchanged a cord of wood worth 7 dollars for a ton of coal worth five dollars; how much did I lose?

15. Two boys went nutting; one brought home 5 pecks and the other 4 pecks; how many pecks did both bring?

16. Frank paid 4 shillings for a pair of doves and 2 shillings for oats to feed them; how much money did he spend for them?

17. Emma, Hattie, and Lucy have each 4 dolls; how many dolls have they all?

18. Count by 4's from 1 to 61.

19. What will it cost to ride 4 miles on the cars, the fare being 2 cents a mile?

20. How many times will the hands of a clock go from XII to VI in 4 days?

DIVISION TABLE. *5's and review.*

"Oft repeated, long remembered."

To be recited "5 in 25 five times," "2 in 6 three times," etc.

```
    a                b                c
25  6  10       16  3  10        9  8  12
 5  2   5        4  3   2        3  4   3
---------       ---------       ----------
 5  3   2        4  1   5        3  2   4

       d                        e
    2  15  20               6  8  4
    2   5   4               3  2  4
    ----------             ---------
    1   3   5               2  4  1

         f                       g
    20   4  15              12   5
     5   2   3               4   5
    ----------              -------
     4   2   5               3   1
```

"5 in 25 five times" may be written 25÷5=5, and is read "25 divided by 5 equals 5."

1. At 3 cents each how many lemons can you buy for 12 cents?

See Teachers Edition. pp. 63, 73.

· Solution: If *one* lemon costs 3 cents, for 12 cents you can buy as many lemons as 3 is contained times in 12, or 4. If preferred, use the following: As many as there are 3's in 12, or 4.

2. 10 cents will buy how many marbles at 2 cents each?

3. Joseph spent 12 cents for oranges, paying 4 cents for each orange; how many did he buy?

4. A boy sold a pair of rabbits for 25 cents; how many oranges at 5 cents each can he buy with the money?

5. Mr. Brown paid a boy 12 shillings for work, at the rate of 2 shillings a day; how many days had the boy worked?

6. Lottie spent 16 cents for candy; she gave 4 cents an ounce; how many ounces did she buy?

7. At 3 cents each how many marbles can Edward buy for 15 cents?

8. Mary's brothers gave her 16 cents, each giving her 4 cents; how many brothers had she?

9. Harry has 8 dollars in the bank; his father has put 2 dollars there for him each birthday; how many birthdays has he seen?

10. If one pineapple costs 2 shillings, how many can you buy for 10 shillings?

11. How many ink wells at 3 cents each can you buy for 9 cents?

12. If the fare on the cars is 4 cents to a certain village, how much is the fare for both ways?

13. Eight little girls were in the woods looking for violets; only 3 girls found any; how many found none?

14. How many quarts in 3 gallons?

15. How many pailfuls of beans will it take to fill an eight-quart basket, if each pail holds 2 quarts?

16. Fred has 5 apples, John has one, and Harry has 3; how many have all?

Read the following numbers: (17.) 321468; (18.) 108320; (19.) 716381; (20.) 400750; (21.) 604025; (22.) 700006; (23.) 800000; (24.) 70016; (25.) 215000; (26.) 380500; (27.) 50000.

It is an excellent exercise for the class to write the foregoing numbers in words.

Write in Arabic:

28. Forty-nine thousand seven hundred sixty.

29. Ten thousand ninety.

30. Three hundred seventeen thousand nine hundred thirty-one.

31. Nine hundred thousand one hundred one.

32. Four hundred thousand forty.

33. Six hundred thousand.
34. Two hundred ninety-one thousand five.
35. Thirty thousand ten.
36. Add 32, 21, 13, 3, 32, 22, 12, 33, 32, 13, 23, and 2.
37. Add 3, 31, 23, 3, 33, 12, 2, 33, 21, 32, 22, 33, and 23.
38. Add 33, 21, 23, 31, 12, 3, 23, 13, 32, 20, 32, and 12.
39. Add 21, 33, 3, 12, 30, 23, 33, 21, 13, 33, 21, and 33.
40. Add 32, 3, 23, 33, 13, 21, 23, 33, 30, 20, 12, 2, and 21.
41. Add 23, 33, 2, 30, 33, 13, 11, 32, 23, 33, 21, and 32.
42. Add 43, 21, 4, 14, 32, 23, 42, 34, 24, 42, 33, and 4.
43. Add 4, 23, 42, 21, 3, 44, 43, 23, 31, 14, 23, and 31.

1. What cost 3 books at 4 shillings each?
2. Mary rode in the swing five times, and Jane 4 times; how many times did they both ride?
3. How many marbles at two cents each can you buy for 6 cents?
4. Jane saw five doves on the ground; three of them flew away; how many remained on the ground?

See Teachers' Edition, p. 69.

5. Bessie went to school 5 days, and Mary went 3; how many more days did Bessie go than Mary?

6. How many times can I take 2 marbles from a pile of 8 marbles?

7. A little girl had 20 cents; how many four-cent lead pencils can she buy with her money?

8. A boy walked 2 miles each day for 4 days; how far did he walk?

9. A boy having three five-cent pieces lost two of them; how many cents had he left?

10. A little girl ate 3 buckwheat cakes for breakfast, and her brother ate 4; how many did both eat?

11. Lewis gave 4 boys 4 marbles apiece; how many did he give them all?

12. Jennie spent 5 cents for raisins, 3 cents for candy, and 2 cents for a stick of gum; how much did she spend?

13. On the east side of a house there are 5 windows; 3 of them are open; how many are closed?

14. A good boy carried 8 pails of water for his mother on Monday, and 5 on Tuesday; how many more did he carry on Monday than on Tuesday?

15. How many quarts in 4 gallons?

16. Among how many children can I

divide 15 plums that each may receive 3 plums?

17. There are 16 towns in Ontario County; if you learn the names of 4 of them each day, in how many days will you learn the names of all of them?

18. Charles had 4 marbles and his brother gave him 5; how many had he then?

19. Henry had to stay after school 5 minutes for whispering, and 2 minutes to solve an example; how long did he have to stay?

20. If John gets 5 cents for husking one bushel of corn, how many bushels must he husk to earn 20 cents?

21. If a boy traded a knife worth 10 cents for a top worth 5 cents, how much did he lose?

22. How much will a boy earn in 4 days at 3 shillings a day?

ADDITION TABLE.   *7's and review.*

a				*b*			
6	3	7	5	2	4	6	3
3	4	5	6	7	3	4	5
9	7	12	11	9	7	10	8

See Teachers' Edition, p. 74.

30 FIRST STEPS AMONG FIGURES.

```
         c                        d
  7 .  5   2   4           7   6   3   7
  6    7   3   4           3   5   6   7
 ─────────────────        ─────────────────
 13   12   5   8          10  11   9  14

         e                        f
  5    2   4   6           3   5   2   4
  3    4   5   6           7   4   5   6
 ─────────────────        ─────────────────
  8    6   9  12          10   9   7  10

         g                        h
  6    3   7               5   2   4
  7    3   4               5   6   7
 ──────────────           ──────────────
 13    6  11              10   8  11
```

SUBTRACTION TABLE.   7's and review.

```
         a                        b
 13   12   5   8          10  11   9   7
  6    7   3   4           3   4   6   5
 ─────────────────        ─────────────────
  7    5   2   4           7   7   3   2

         c                        d
 14    9   7  12          11   9   7  10
  7    3   4   5           6   7   3   4
 ─────────────────        ─────────────────
  7    6   3   7           5   2   4   6
```

$$
\begin{array}{cccc}
& & e & \\
8 & 9 & 10 & 13 \\
5 & 4 & 7 & 7 \\
\hline
3 & 5 & 3 & 6
\end{array}
\qquad
\begin{array}{cccc}
& & f & \\
6 & 10 & 6 & 10 \\
3 & 6 & 4 & 5 \\
\hline
3 & 4 & 2 & 5
\end{array}
$$

$$
\begin{array}{ccc}
& g & \\
12 & 9 & 8 \\
6 & 5 & 3 \\
\hline
6 & 4 & 5
\end{array}
\qquad
\begin{array}{ccc}
& h & \\
11 & 8 & 11 \\
5 & 6 & 7 \\
\hline
6 & 2 & 4
\end{array}
$$

MULTIPLICATION TABLE. *7's and review.*

$$
\begin{array}{cccc}
& a & & \\
6 & 3 & 7 & 5 \\
3 & 4 & 5 & 6 \\
\hline
18 & 12 & 35 & 30
\end{array}
\qquad
\begin{array}{cccc}
& b & & \\
2 & 4 & 6 & 3 \\
7 & 3 & 4 & 5 \\
\hline
14 & 12 & 24 & 15
\end{array}
$$

$$
\begin{array}{cccc}
& c & & \\
7 & 5 & 2 & 4 \\
6 & 7 & 3 & 4 \\
\hline
42 & 35 & 6 & 16
\end{array}
\qquad
\begin{array}{cccc}
& d & & \\
7 & 6 & 3 & 7 \\
3 & 5 & 6 & 7 \\
\hline
21 & 30 & 18 & 49
\end{array}
$$

$$
\begin{array}{cccc}
& e & & \\
5 & 2 & 4 & 6 \\
3 & 4 & 5 & 6 \\
\hline
15 & 8 & 20 & 36
\end{array}
\qquad
\begin{array}{cccc}
& & f & \\
3 & 5 & 2 & 4 \\
7 & 4 & 5 & 6 \\
\hline
21 & 20 & 10 & 24
\end{array}
$$

	g			*h*	
6	3	7	5	2	4
7	3	4	5	6	7
42	9	28	25	12	28

DIVISION TABLE.   *7's and review.*

	a				*b*		
49	18	30	21	16	6	35	42
7	6	5	3	4	3	7	6
7	3	6	7	4	2	5	7

	c				*d*		
15	8	20	36	15	24	12	14
3	4	5	6	5	4	3	7
5	2	4	6	3	6	4	2

	e				*f*		
30	35	12	18	10	20	21	24
6	5	4	3	5	4	7	6
5	7	3	6	2	5	3	4

	g			*h*	
42	9	28	25	12	28
7	3	4	5	6	7
6	3	7	5	2	4

See Teachers' Edition, p. 95.

FIRST STEPS AMONG FIGURES.

To the teacher. Read carefully the prefaces to both the Pupils' and the Teachers' Editions, and also the Special Notice which precedes the latter.

Solve the following examples in addition :—

1.	2.	3.	4.	5.	6.
23	33	31	31	32	21
31	23	33	13	13	13
33	31	22	12	33	32
20	20	30	33	21	33
13	13	23	30	12	20
32	33	13	22	32	32
23	32	32	31	23	3
31	23	31	13	13	31
33	31	23	23	32	31
12	33	33	31	23	23

7.	8.	9.	10.
32	33	31	32
23	31	12	23
31	23	33	31
33	33	22	23
21	12	31	13
13	23	23	31
22	31	12	20
33	32	33	33
31	23	32	23
23	33	23	31

Some of these examples are to be given daily. See Teachers' Edition, p. 88.

Solve the following examples in addition :—

11. 32	12. 33	13. 23	14. 33	15. 32
21	21	33	22	23
13	32	31	31	30
33	33	22	23	33
30	13	33	20	22
2	32	21	32	31
23	3	13	31	23
31	12	21	23	33
23	30	32	33	20
32	21	33	31	3
33	33	13	23	31
23	32	33	33	33

16. 32	17. 33	18. 21	19. 31
23	22	33	23
23	30	32	31
30	21	23	23
23	13	33	30
31	32	22	33
22	33	31	21
32	22	23	13
21	31	32	32
32	23	13	32
33	22	33	23
23	31	31	32

Some of these examples are to be given daily. See Teachers' Edition, p. 89.

FIRST STEPS AMONG FIGURES.

Solve the following examples in addition :—

20.	21.	22.	23.
32	33	31	23
13	21	23	31
23	33	33	23
31	32	31	33
33	22	22	30
23	23	32	23
31	31	23	12
21	32	21	33
12	13	32	23
32	21	33	32
33	23	32	33
32	33	33	23

24.	25.	26.	27.
33	23	31	13
21	31	23	33
30	22	32	21
13	30	21	32
23	3	13	33
31	32	33	23
30	52	33	33
23	23	20	20
31	33	13	12
12	31	32	33
23	32	33	32
32	33	23	12

Some of these examples are to be given daily.

FIRST STEPS AMONG FIGURES.

Solve the following examples in addition:—

28. 32	29. 33	30. 21	31. 33
20	21	32	23
33	32	23	30
23	33	33	22
12	21	21	12
33	13	23	33
31	33	33	21
22	33	13	21
33	22	23	13
32	31	33	33
21	33	21	30
33	23	32	22

32. 30	33. 21	34. 23
23	33	32
12	12	13
31	23	23
33	32	32
22	31	33
32	32	23
23	20	32
31	11	30
33	32	31
22	23	23
31	33	22

Some of these examples are to be given daily.

FIRST STEPS AMONG FIGURES. 37

After these examples have all been solved once, unless the pupils are very ready in adding, let them commence at the first example and solve them all again It is of the utmost importance that the pupil should learn to add quickly and correctly, for in practical life he will use addition a dozen times where he will use fractions once.

*Examples in Subtraction.

35. 75986	36. 96897	37. 68795
54234	72044	5250

38. 69587	39. 95786	40. 68795
47043	2053	6203

41. 86975	42. 79586	43. 68579
20324	2102	40234

44. 76859	45. 97586	46. 68579
4506	35430	5345

*Examples in Multiplication.

1. 23,103	2. 30,231	3. 13,202	4. 21,032
2	3	3	2

See Teachers' Edition, p. 91.

5. 31,023	6. 12,320	7. 20,123	8. 23,132
3	2	3	3

9. 31,420	10. 32,301	11. 30,423	12. 23,123
2	3	2	3

*Examples in Division.

13. 2)64,682   14. 3)93,609   15. 2)20,486

16. 2)40,826   17. 3)39,069

18. 2)82,604   19. 2)28,064

20. 3)90,639   21. 2)26,048

22. 2)80,264   23. 3)60,936

See Teachers' Edition, p. 92.

Examples in Addition.

24. 342	25. 234	26. 434
231	423	244
423	344	342
443	244	444
334	432	322
234	343	434
422	443	324

27. 213	28. 444	29. 431
432	231	242
444	443	344
344	232	423
231	424	344
414	344	332
342	231	421

30. 34	31. 243	32. 4
442	424	342
321	331	234
434	212	23
444	443	341
343	344	422
212	424	444
243	341	341
444	234	213

* The teacher is to show the pupils how to solve such examples.

33.
| 134 |
| 423 |
| 344 |
| 103 |
| 422 |
| 344 |
| 441 |
| 343 |
| 413 |

34.
| 324 |
| 431 |
| 344 |
| 223 |
| 431 |
| 444 |
| 324 |
| 213 |
| 444 |

35.
| 432 |
| 341 |
| 223 |
| 312 |
| 434 |
| 244 |
| 443 |
| 332 |
| 423 |

36.
| 2342 |
| 4324 |
| 3432 |
| 2343 |
| 3423 |
| 2343 |
| 3434 |
| 4232 |
| 1443 |
| 3234 |
| 4342 |

37.
| 4323 |
| 2433 |
| 4344 |
| 4224 |
| 3433 |
| 1234 |
| 4322 |
| 2443 |
| 3244 |
| 4432 |
| 2343 |

38.
| 3423 |
| 2342 |
| 4214 |
| 2343 |
| 4242 |
| 1234 |
| 4321 |
| 3244 |
| 3423 |
| 4343 |
| 2432 |

See Teachers' Edition, p. 93.

FIRST STEPS AMONG FIGURES.

39. 3424
     4132
     3434
     1234
     4343
     2432
     1243
     3324
     4432
     3243
     4432

40. 4234
     3442
     2323
     4342
     1234
     3421
     4343
     2422
     3234
     4343
     2342

41. 343
     231
     444
     332
     243
     444
     423
     411
     344
     432
     243
     444
     321

42. 341
     234
     443
     444
     321
     213
     442
     344
     431
     342
     214
     404
     341

43. 431
     343
     234
     321
     443
     444
     324
     242
     441
     304
     323
     444
     321

44. 213	45. 341	46. 423
441	232	344
334	444	231
234	323	444
443	213	324
342	441	431
134	344	344
423	422	444
444	343	321
324	243	423
443	412	344
322	333	442
203	431	322

47. 3424	48. 4234	49. 4324
4323	3342	3432
3243	4433	4243
4434	1234	2324
2332	3243	4433
3443	4324	3442
4324	2442	2334
2423	3334	4343
4244	4234	2434
3423	2442	2343
3342	3223	4123
4324	4324	3434
3423	3443	4442
2342	4321	3223

See Teachers' Edition, p. 94.

50.	3423	51.	3443
	3244		4234
	4334		2343
	2433		4324
	4242		4432
	3423		2444
	4344		4321
	1234		3443
	4422		4242
	4342		4433
	3423		3444
	4444		2344
	3444		4323
	2331		3434

1. Richard has 7 cents and Oliver has 6 cents, how many have both?
2. Cora bought 5 sticks of candy and Hattie 4, how many did both buy?
3. Herman had 11 cents, but he lost 3 of them, how many had he then?
4. Herbert had 7 marbles, he found 5 more, how many had he then?
5. Ella had 8 new needles, she broke 5 of them, how many whole ones had she then?
6. Clara solved 12 examples while Kitty solved 7, how many more did Clara solve than Kitty?

See Teachers' Edition, p. 97.

7. Anna had 10 oranges, she gave away 4 of them, how many had she left?

8. Frank had 4 pencils, he bought 3 more; how many had he then?

9. What cost 7 lemons at 4 cents each?

10. If an orange cost 5 cents, what will 3 oranges cost?

11. If an orange cost 6 cents, and a lemon cost 5 cents, what will both cost?

12. How many quarts in 3 gallons?

13. Henry had 13 apples, he gave his brother 7 of them; how many had he then?

14. At 5 shillings a bushel, what will 6 bushels of apples cost?

15. What cost 7 pencils at 3 cents each?

16. A boy walked 4 miles in the morning and 3 miles in the afternoon. How far did he walk?

17. Samuel has 7 cents in one pocket and 5 cents in another, how many in both?

18. Carlos had 9 cents in his pocket, he lost 4 of them through a hole in his pocket, how many had he then?

19. Mary spelled 8 words correctly and Emma spelled 6 words correctly, how many more did Mary spell correctly than Emma?

20. At 3 cents each, how many marbles can you buy for 15 cents?

21. 12 boys were sliding on the ice, 7 of them fell, how many remained standing?

See Teachers' Edition, p. 98.

22. If a boy earn 7 cents a day, how many days will it take him to earn 42 cents?

23. If one lead-pencil cost 6 cents, how many can be bought for 30 cents?

24. James had 12 cents, he spent 5 of them and lost 3 more, how many had he left?

Solution: He disposed of the sum of 5 cents and 3 cents or 8 cents. He had left the difference between 8 cents and 12 cents or 4 cents; or, if he spent 5 cents and lost 3,· he disposed of the sum of 5 cents and 3 cents or 8 cents. If he had 12 cents and disposed of 8 cents he had left the difference between 12 cents and 8 cents or 4 cents.

25. Sarah had 10 needles; she gave 3 of them to Nellie and 4 of them to Martha. How many had Sarah left?

26 Herbert had 6 cents, he earned 7 cents and spent 5 cents; how many had he then?

27. Joel had 7 cents, he earned 5 cents and found 4 cents; how many had he then?

28 How many days will it take Walter to walk 28 miles, if he walks 4 miles each day?

29. An orange cost 6 cents and a cocoanut 3 times as many. How much did the cocoanut cost?

2*

30. Mary had 11 cents, she lost 5, and earned enough to make her number 9. How many did she earn?

31. There are 4 columns in John's spelling lesson and 5 words in each column. How many words in his lesson?

32. Some boys are out flying kites, the wind blows down two kites and 7 less 3 remain. How many at first in the air?

33. A boy hoed corn for 4 cents a row and earned 24 cents. How many rows did he hoe?

34. A man can walk a mile in 10 minutes, he starts from his home and walks to town in 5 minutes. How far from town does he live?

35. I had 10 cents. I bought 2 two-cent stamps, and gave 4 cents to a poor little boy for bread; how many left for candy?

36. There are 9 ten o'clock scholars this morning and one more than one third of them left their books at home. How many of them brought their books?

37. Willie had 13 cents and he spent 6 of them for candy; how many had he left?

38. How many sponges at 6 cents each may be bought for 36 cents?

39. What cost a pencil at 7 cents and a pint of peanuts at 5 cents?

40. What cost 7 pencils at 5 cents each? Count by 5's from 2 to 62.

Review counting by 5's from 1 and 5 to 61 and 60. Review counting by 4's.

Count by 5's from 3 to 63.

Count by 5's from 4 to 64.

1. 345	2. 35	3. 405
452	454	352
553	542	544
245	135	255
523	523	533
345	454	441
521	542	355
453	5	523
45	352	344
352	534	555
424	435	434

4. 345	5. 505	6. 305
532	453	543
454	344	444
325	535	354
543	413	532
355	553	445
324	435	354
513	214	543
454	543	455
545	345	544
353	553	353

Examples in Subtraction.

7. 79,687
   64,252

8. 96,487
   31,343

9. 79,689
   43,264

10. 687,985
    34,532

11. 795,869
    43,543

12. 47,685
    5,232

13. 96,758
    54,428

14. 5,879
    2,343

15. 764,358
    201,223

16. 764,037
    23,010

17. 423,703
    12,400

18. 647,094
    24,070

19. 74,687
    3,053

20. 74,687
    30,435

21. 695,047
    252,004

22. 523,012
    46,566

23. 613,021
    56,554

24. 431,024
    55,357

See Teachers' Edition, p. 101.

25. 634,210
       56,643
	26. 820,132
	       53,657
		27. 431,201
		       54,464

28. 942,031
       66,466
	29. 731,203
	       64,636
		30. 841,310
		       63,653

31. 312,403
       45,257
	32. 420,314
	       54,652
		33. 341,531
		       34,354

34. 213,042
       36,527
	35. 314,253
	       51,646
		36. 453,621
		       32,365

37. 324,102
       46,637
	38. 231,430
	       12,257
		39. 425,301
		       51,625

40. 430,221
       52,147
	41. 320,413
	       63,257
		42. 534,102
		       62,637

43. 534,210
       16,436
	44. 130,241
	       53,725
		45. 342,013
		       30,157

46. 703,524
       20,352
	47. 423,102
	       47,056
		48. 624,130
		       61,563

FIRST STEPS AMONG FIGURES.

If more practice is desired at this stage the foregoing examples may be reviewed. Count by 6's from 6 to 60.

1. 4532	2. 3541	3. 5423
3215	4325	4352
5453	5432	3544
3344	3254	5435
2535	5545	4334
4253	3251	2453
5432	4314	3545
2354	2443	4334
4235	5425	5442
5542	4543	2355

4. 3524	5. 4325
4352	3552
3445	5445
5334	3453
4523	4524
5452	5335
3445	3452
4334	5344
5523	2325
4354	4553

See Teachers' Edition, p. 103.

FIRST STEPS AMONG FIGURES.   51

6. 3542	7. 245	8. 5342
4354	4532	3534
2435	2454	4253
5043	5321	5425
3530	3543	343
354	4254	4534
4245	5435	2345
5432	2353	5234
3554	540	4553
4322	3254	3425
2453	4342	5342
5245	5535	2435

9. 3045	10. 3452
4532	5544
5450	2345
3345	4532
2534	5453
350	1324
4003	5045
5435	3453
3542	4234
4354	4553
5425	5345
4543	3522

Multiply:

11. 2,314
     2

12. 4,231
     2

13. 2,014
     2

14. 30,124
      2

15. 21,302
      3

16. 34,201
      2

17. 23,041
      2

The teacher should show the pupils how to solve the following examples:

18. 35,246
      2

19. 42,356
      3

20. 46,352
      2

21. 35,642
      3

22. 36,452
      3

23. 26,453
      3

24. 46,352
      2

25. 64,526
      3

26. 53,624
      4

27. 64,524
      3

See Teachers' Edition, p. 104.

28. 53,426
       4
	29. 64,536
	       4
		30. 35,642
		       2

31. 46,352
       3
	32. 36,426
	       4
		33. 35,246
		       3

34. 36,452
       6
	35. 36,546
	       6
		36. 64,526
		       6

37. 46,352
       5
	38. 63,524
	       4
		39. 35,042
		       4

40. 40,536
       5
	41. 30,462
	       6
		42. 46,035
		       5

43. 26,304
       6
	44. 25,036
	       6
		45. 52,064
		       6

Count by 6's from 1 to 61.
Count by 6's from 2 to 62.
Count by 6's from 3 to 63.
Count by 6's from 4 to 64.
Count by 6's from 5 to 65.

Addition Table. *8's and review.*

```
      a                  b                    c
7 4 8 5 2         6 3 7 4 8  5      2 6 3  7
3 4 5 6 7         8 3   4 5  6      7 8 3 4 5
─────────         ──────────────    ─────────────
10 8 13 11 9      14 6 11 9 14      12 10 9 7 12

      d                  e                    f
3 7 4 8 5         2 6 3 7 4          8 5 2  6 3
5 6 7 8 3         4 5 6 7 8          3 4 5  6 7
─────────         ─────────          ─────────────
8 13 11 16 8      6 11 9 14 12       11 9 7 12 10

      g                  h                    i
7 4 8 5 2         6 3 4 8             5 2 6
8 3 4 5 6         7 8 6 7             8 3 4
─────────         ─────────           ─────────
15 7 12 10 8      13 11 10 15         13 5 10
```

Subtraction Table. *8's and review.*

```
      a                   b                   c
10 12 7 15 8       13 11 9 7 12        10 11 12 14
 5  4 3  8 6        7  8 4 5  6         7  3  8  7
──────────────     ──────────────      ─────────────
 5  8 4  7 2        6  3 5 2  6         3  8  4  7

      d                   e                   f
9.11 6 8 10        14 5 13 8 13        11 9 14 11 16
6  5 4 3  4         6 3  8 5  6         7 7  8  4  8
──────────────     ──────────────      ─────────────
3  6 2 5  6         8 2  5 3  7         4 2  6  7  8

      g                   h                   i
9 6 12 13 8        10 10 9 7           12 10 15 11
5 3  7  5 4         3  8 3 4            5  6  7  6
──────────────     ──────────────      ─────────────
4 3  5  8 4         7  2 6 3            7  4  8  5
```

See Teachers' Edition, p. 105.

FIRST STEPS AMONG FIGURES. 55

MULTIPLICATION TABLE. *8's and review.*

```
       a                  b                  c
  7  4  8  5  2      6  3  7  4  8      5  2  6  3
  3  4  5  6  7      8  3  4  5  6      7  8  3  4
 ─────────────      ─────────────      ─────────────
 21 16 40 30 14    48  9 28 20 48     35 16 18 12

       d                  e                  f
  7  3  7  4  8      5  2  6  3  7      4  8  5  2
  5  5  6  7  8      3  4  5  6  7      8  3  4  5
 ─────────────      ─────────────      ─────────────
 35 15 42 28 64    15  8 30 18 49     32 24 20 10

       g                  h                  i
  6  3  7  4  8      5  2  6  3  4      8  5  2  6
  6  7  8  3  4      5  6  7  8  6      7  8  3  4
 ─────────────      ─────────────      ─────────────
 36 21 56 12 32    25 12 42 24 24     56 40  6 24
```

DIVISION TABLE.   *8's and review.*

```
         a                    b                   c
  28  9 48 14 30       40 16 21 20 48     35 16 18 12
   4  3  8  7  6        5  4  3  5  6      7  8  3  4
 ──────────────       ──────────────      ─────────────
   7  3  6  2  5        8  4  7  4  8      5  2  6  3

         d                    e                   f
  35 24 56 42 15       24  6 40 28 64     15  8 21 36
   5  6  7  6  5        4  3  8  7  8      3  4  7  6
 ──────────────       ──────────────      ─────────────
   7  4  8  7  3        6  2  5  4  8      5  2  3  6

         g                    h                   i
  10 20 24 32 49       18 30 32 25 12     42 24 56 12
   5  4  3  8  7        6  5  4  5  6      7  8  8  3
 ──────────────       ──────────────      ─────────────
   2  5  8  4  7        3  6  8  5  2      6  3  7  4
```

See Teachers' Edition, p. 105.

1. George paid 15 cents for a knife, and after breaking it sold it for 8 cents. How many cents did he lose?

2. Lewis bought a reader for 6 shillings and an arithmetic for 4 shillings; how much money did he spend for both?

3. Anna is 13 years old and Mary is 5 years old. How many years older is Anna than Mary?

4. John had 8 cents; he lost 5 of them, how many had he left?

5. What cost 4 books at 8 shillings each?

6. How many oranges at 5 cents each can you buy for 40 cents?

7. Walter spent 16 cents for pears at 2 cents each; how many pears did he get?

8. Jane had 15 needles; she lost 3 of them and broke 5; how many had she left?

9. Byron rode down hill 4 times one afternoon and his father twice as many times. How many times did Byron's father ride down hill?

10 William had 26 cents; he lost two of them and spent the rest for marbles at 3 cents each. How many marbles did he buy?

11. Charles was sent to the store to buy 6 spools of thread at 8 cents a spool. He took 50 cents with him, how much change should he take home?

See Teachers' Edition, p. 108.

FIRST STEPS AMONG FIGURES. 57

12. A boy having 8 cents earned 5 cents and his sister gave him 2 cents, how many had he then?

13. A boy set 2 traps in the woods; 2 rabbits went into one trap and twice as many went into another. How many went into both?

14 A gun carriage has four wheels; how many wheels have 7 gun carriages?

15. If it takes 6 horses to draw one cannon, how many horses will draw 8 cannons?

16 There are a sergeant and 6 privates at one picket post, and a corporal and 4 privates at another. How many soldiers at both?

17. If 3 oranges cost 12 cents what will one orange cost?

Solution: If *three* oranges cost 12 cents, *one* orange will cost one-third of 12 cents or 4 cents

We get $\frac{1}{3}$ of a number by dividing it by 3, $\frac{1}{4}$ of a number by dividing it by 4, etc.

What is one-fourth of 24?

What is one-third of 21?

What is one sixth of 42? One-fifth of 30? One-fourth of 32? One-seventh of 56? One-half of 14? One-sixth of 18?

18. If 8 horses eat 24 bushels of oats in a week, how many bushels will one horse eat?

19. If 42 cents is the price of 7 marbles, what is the price of one marble?

20. 4 boys have 28 cents, and each have an equal number. How many cents has each boy?

21. A farmer has 18 pigs in 3 pens and the same number in each pen. How many pigs in one of the pens?

22. If 8 pears cost 16 cents, what cost 1 pear?

23. At 4 cents each, how many lemons can be bought for 28 cents?

24. If three tops cost 15 cents, what cost 1 top?

25. What cost 1 bushel of oats, if 7 bushels cost 35 shillings?

26. 12 dollars will buy how many birds at 2 dollars each?

27. If 5 hens cost 20 shillings what will 1 hen cost?

28. How many peaches at 2 cents each can you buy for 16 cents?

FIRST STEPS AMONG FIGURES.

1. 5304	2. 453	3. 5243
4243	304	4524
3524	5035	3452
5452	4540	5335
535	3235	3543
4343	2424	4254
3425	5353	5435
5254	4545	3542
2542	3424	5354
4335	4353	4435
3454	2532	3423
5543	4245	5354
4325	5454	3543

4. 4305	5. 3544
3450	5434
5234	4355
2543	3243
5455	5424
3213	4552
4543	2345
2355	5433
5432	2253
4343	4534
2524	3445
3455	2343
3532	4554

See Teachers' Edition p. 110

6. 4305	7. 2345	8. 5435
3453	5334	554
435	4243	3443
5044	3545	5545
3453	5434	4353
4545	3552	5534
5432	3345	3444
3354	5453	5554
3545	4245	2345
34	3524	4235
5432	4305	5453
4355	5353	3544
4544	3445	5355

9. 3452	10. 3452
2345	5435
5443	4544
4534	3353
4455	5445
5243	3234
3524	5544
4355	4353
5533	2455
4432	4434
5454	5554
3345	5432
5432	3245

FIRST STEPS AMONG FIGURES. 61

Count by 6's from 2, 3, 4 and 5, to 62, 63, 64 and 65.

Read the following numbers:
1. 750000748
2. 90000047
3. 680000740
4. 700746000
5. 12003000
6. 750908716
7. 801000071
8. 679374819
9. 715016390
10. 900060000
11. 7000000
12. 800000005

Addition :

Finding how many units there are in two or more numbers and expressing them in one number is called *addition*.

The number found by addition is called the *sum* or *amount*.

Proof. Add the columns both upward and downward, and if the results agree they are probably correct.

11. Write in Arabic fifteen mil. ten th. ninety.

12. Write in Arabic three hun. fifty mil. nine hun. th.

13. Write in Arabic eight mil. five hun. forty-three th. seven.

14. Write in Arabic one hun. mil. six hun. thirty-two.

15. Write in Arabic seventy-five mil. three hun th.

16. Write in Arabic fifty mil fifty th. fifty.
17. Write in Arabic two hun. mil. sixty th. four.
18. Write in Arabic one hun. five mil. one hun five.
19. Write in Arabic one hun. nineteen mil. forty th.
20. Write in Arabic three hun. eight mil. thirteen th. two hun. eighty-one.
21. Write in Arabic twenty th.
22. Write in Arabic fifty mil. fifty.
23. Write in Arabic three th. two hun. forty-five.
24. Write in Arabic nineteen mil. five hun. th.
25. Write in Arabic eight mil. three hun.
26. Write in Arabic four hun. mil.
27. Write in Arabic seven hun. sixty-one mil. five hun. sixteen th. twenty.
28. Write in Arabic six hun th.
29. Write in Arabic fifty th. forty.
30. Write in Arabic three hun. eight.
31. Write in Arabic nine th.
32. Write in Arabic six hun. mil. five. hun.
33. Write in Arabic sixty mile, four hun. th. three hun.
34. Write in Arabic three mil. fifteen th. thirty.

See Teachers' Edition, p. 114.

FIRST STEPS AMONG FIGURES. 63

*35. Write in words 13,004,020.
36. Write in words 300,216,000.
37. Write in words 232,341,519.
38. Write in Roman four hun. sixty-three.
39. Write in Roman eight hun. forty-four.
40. Write in Arabic DCLXXVII.
41. Write in Roman two hun. eighty-nine.
42. Write in Arabic XCVIII.
43. Write in Arabic CDXI.
44. Write in Roman seven hun. nineteen.

Subtraction:

Taking one number from another number is called subtraction.

Remainder or difference.

The number found by taking one number from another is called the *difference* or *remainder.*

In subtraction the number to be subtracted is called the *subtrahend*, and the number it is subtracted from is called the *minuend*. The result in subtraction is called the *difference* or *remainder*.

PROOF. Add the remainder and the subtrahend, and if the result equals the minuend the work is probably correct.

The pupils should often be required to

---

* Notice that such compound words as seventy-five, forty-one, sixty-nine, etc., require a hyphen.

FIRST STEPS AMONG FIGURES.

write the name of each number in examples in subtraction, as follows:

321 Minuend.
45 Subtrahend.

276 Remainder or dif.

Subtraction.

(1)	(2.)	(3.)
8,342,053	73,520,031	81,400,253
474,377	243,075	2,302,675

(4.)	(5.)	(6.)
71,420,035	93,520,014	4,531,024
241,068	5,700,043	765,337

(7.)	(8.)	(9.)
74,200,352	6,314,025	73,500,241
5,402,568	565,068	212,376

(10.)	(11.)	(12.)
94,002,531	81,350,024	34,200,156
7,006,763	576,546	7,620,478

See Teachers' Edition, p. 118.

(13)
7,352,034
181,376

(14.)
95,300,421
420,376

(15)
863,005,241
2,307,374

(16.)
93,510,042
46,350,357

(17.)
35,200,416
2,540,348

(18.)
7,300,425
602,557

(19.)
83,001,425
5,005,743

(20.)
95,320,041
4,460,474

(21.)
85,241,300
7,534,720

(22.)
94,300,521—570,376=?

(23.)
42,530,014—6,765,068=?

(24)
71,300,524—1,000,677=?

(25.)
82,400,153—4,740,075=?

The teacher will choose between the two following sets of definitions:

Multiplication.

A short method of adding equal numbers is called *multiplication*.

Multiplicand.

One of the equal numbers is called the *multiplicand*.

Multiplier.

The number which shows how many of the equal numbers are used in adding is called the *multiplier*.

In the example 542 + 542 + 542, 542 is the multiplicand and 3 is the multiplier, and we solve the example as follows:

$$\begin{array}{r} 542 \text{ Multiplicand,} \\ 3 \text{ Multiplier,} \\ \hline 1{,}626 \text{ Product.} \end{array}$$

Or the following definitions may be used:

Multiplication.

Taking a number a certain number of times is called *multiplication*.

Multiplicand.

The number taken, (or multiplied,) is called the *multiplicand*.

FIRST STEPS AMONG FIGURES. 67

Multiplier.

The number by which we multiply, (or which shows how many times the multiplicand is taken,) is called the *multiplier.*

Product.

The result of the multiplication is called the *product.*

Proof.

I. Multiply the multiplier by the multiplicand, and if the product equals the product first obtained, the work is probably correct.

II. Or divide the product by the multiplicand, and if the work is correct the result will be the multiplier, or divide by the multiplier and get the multiplicand.

Multiplication.

1. 35,246
         7

2. 42,356
         7

3. 35,642
         6

4. 57,463
         7

5. 35,642
         7

6. 35,246
         8

7. 57,463
         8

8. 35,246
         6

9. 46,058
         8

See Teachers' Edition, p. 119.

10. 685,037	11. 485,067	12. 857,046
8	7	6

13. 637,058	14. 364,078
4	7

15. $3,640,758 \times 8 = ?$ 16. $7,583,640 \times 7 = ?$

When there is more than one figure in the number by which we multiply, the right hand figure of the result is placed under the figure we multiply by.

17. 46,352	18. 56,342	19. 2,635
32	23	32

20. $364,526 \times 43 = ?$
21. $35,264 \times 45 = ?$
22. $42,563 \times 54 = ?$
23. $534,652 \times 46 = ?$
24. $645,362 \times 64 = ?$
25. $63,527 \times 76 = ?$
26. $4675 \times 67 = ?$
27. $57,463 \times 74 = ?$

FIRST STEPS AMONG FIGURES.

28. 357,426 × 63 = ?
29. 47,563 × 67 = ?
30. 63,527 × 54 = ?
31. 352,746 × 34 = ?
32. 475,063 × 57 = ?
33. 630,574 × 76 = ?
34. 57,463 × 35 = ?
35. 357,426 × 47 = ?

Division.

Finding how many times one number is contained in another is called *division.*

Dividend.

The number which contains the other is called the *dividend.* (If preferred, the number divided is called the *dividend.*)

Divisor.

The number which is contained in the other is called the *divisor.* (If preferred, the number by which we divide is called the *divisor.*)

Quotient.

The number found by dividing is called the *quotient.*

Proof.

Multiply the quotient by the divisor, and add the remainder if there be one; if the work is correct the result should equal the dividend.

Divisor, 7)4321 Dividend.

Quotient, 617—2 Remainder.

Short Division.

(1.)	(2.)	(3.)
2)6,846	3)612,921	4)3,281,224

(4.)	(5.)	(6.)
3)9,182,715	4)2,082,836	5)5,304,535

(7.)	(8.)	(9.)
3)24,152,721	2)128,166,414	4)836,284,820

(10.)	(11.)	(12.)
3)21,186,912	4)2,032,828	2)106,416,818

See Teachers' Edition, p. 120.

With remainders. *

13. $7,372 \div 3 = ?$
14. $14,173 \div 4 = ?$
15. $16,427 \div 3 = ?$
16. $823,975 \div 3 = ?$
17. $2,182,939 \div 4 = ?$
18. $1,327,178 \div 5 = ?$
19. $954,546 \div 4 = ?$
20. $5,663,032 \div 3 = ?$
21. $26,173,385 \div 4 = ?$
22. $273,823 \div 5 = ?$
23. $925,588 \div 6 = ?$
24. $257,429 \div 4 = ?$
25. $3,741,254 \div 6 = ?$
26. $376,233 \div 5 = ?$
27. $3,804,168 \div 7 = ?$
28. $4,873,290 \div 5 = ?$
29. $4,497,190 \div 7 = ?$
30. $3,880,916 \div 6 = ?$
31. $3,146,153 \div 4 = ?$
32. $172,448 \div 7 = ?$
33. $20,740,965 \div 6 = ?$

Count by 7's from 7 to 42.

* Teach the division series with remainders in Teachers' Edition, p. 107, before solving these examples.

Addition Table. 9's and review.

```
      a              b               c
 6  9  5  8  4 | 7  3  6  9  5 | 8  4  7  3  6
 4  5  6  7  8 | 9  4  5  6  7 | 8  9  4  5  6
─────────────── ─────────────── ───────────────
10 14 11 15 12 |16  7 11 15 12 |16 13 11  8 12

      d              e               f
 9  5  8  4  7 | 3  6  9  5  8 | 2  7  3  6  9
 7  8  9  4  5 | 6  7  8  9  4 | 9  6  7  8  9
─────────────── ─────────────── ───────────────
16 13 17  8 12 | 9 13 17 14 12 |11 13 10 14 18

         g                    h
  5  8  4  7  3  6 |  9  5  8  4  7  3
  4  5  6  7  8  9 |  4  5  6  7  8  9
 ────────────────── ──────────────────
  9 13 10 14 11 15 | 13 10 14 11 15 12
```

Subtraction Table. 9's and review.

```
       a              b              c
12 15 11 14 11 |15 13 10 14 10 |13  9 13 10
 9  8  7  6  8 | 9  4  5  7  6 | 5  4  6  7
─────────────── ─────────────── ───────────
 3  7  4  8  3 | 6  9  5  7  4 | 8  5  7  3

       d              e              f
14 18  9 12 14 |17  8 12  9 13 |17 13 16 12  8
 8  9  5  4  9 | 8  4  5  6  7 | 9  8  7  6  5
─────────────── ─────────────── ──────────────
 6  9  4  8  5 | 9  4  7  3  6 | 8  5  9  6  3

          g                     h
11 13 16 12 15 11 |  7 16 10 14 11 15 12
 4  9  8  7  6  5 |  4  9  4  5  6  7  8
─────────────────── ──────────────────────
 7  4  8  5  9  6 |  3  7  6  9  5  8  4
```

See Teachers' Edition, p. 123.

FIRST STEPS AMONG FIGURES. 73

MULTIPLICATION TABLE. *9's and review.*

```
        a                    b                    c
   6  9  5  8  4      7  3  6  9  5      8  4  7  3
   4  5  6  7  8      9  4  5  6  7      8  9  4  5
  ─────────────      ─────────────      ─────────────
  24 45 30 56 32    63 12 30 54 35    64 36 28 15

        d                    e                    f
   6  9  5  8  4      7  3  6  9  5      8  2  7  3
   6  7  8  9  4      5  6  7  8  9      4  9  6  7
  ─────────────      ─────────────      ─────────────
  36 63 40 72 16    35 18 42 72 45    32 18 42 21

             g                          h
   6  9  5  8  4  7  3      6  9  5  8  4  7  3
   8  9  4  5  6  7  8      9  4  5  6  7  8  9
  ────────────────────     ────────────────────
  48 81 20 40 24 49 24     54 36 25 48 28 56 27
```

DIVISION TABLE.   *9's and review.*

```
         a                     b                    c
  32 56 30 45 24      63 12 30 54 35      64 36 28 18
   8  7  6  5  4       9  4  5  6  7       8  9  4  9
  ─────────────      ─────────────      ─────────────
   4  8  5  9  6       7  3  6  9  5       8  4  7  2

         d                     e                    f
  42 18 35 16 72      40 63 36 72 45     32 20 42 21
   7  6  5  4  9       8  7  6  8  9      4  5  6  7
  ─────────────      ─────────────      ─────────────
   6  3  7  4  8       5  9  6  9  5      8  4  7  3

              g                            h
  48 81 49 24 40 18 54     36 24 48 28 56 27 25
   8  9  7  6  5  2  9      4  8  6  7  8  9  5
  ────────────────────     ────────────────────
   6  9  7  4  8  9  6      9  3  8  4  7  3  5
```

See Teachers' Edition, p. 124.

1. In a school there are 8 recitation rooms, each room has 7 benches; how many benches in these rooms?

2. There were 60 freight cars on a road and 6 of them were destroyed in a collision; how many remained?

3. If a man can draw 7 loads of sand in 1 day, how many days will it take him to draw 42 loads?

4 In a class-room there are 6 seats, and each seat will hold 8 pupils; how many pupils can be seated in the room?

5. Six sheep were put into a flock containing 87; how many then in the flock?

6. Pineapples are 6 cents each; how many can be bought for 48 cents?

Count by 7's from 7 to 70.

7. If Nellie pays 5 cents for candy and has 8 cents left, how many cents had she at first?

8. How many legs have 7 cats?

9. How many quarts in 8 gallons?

10. How many quarts in 12 pints?

11. If John buys some candy for 18 cents, some peanuts for 7 cents and an orange for 6 cents, how many cents does he spend?

See Teachers' Edition, p. 123.

12 *If 8 sheep cost $32, what will one sheep cost?

13 How many fingers and thumbs have 6 boys?

14. Willie bought 2 pounds of crackers at 8 cents a pound, and half a pound of cheese at 14 cents a pound; how much money did he spend?

15. Harry had a ten-cent piece, a five-cent piece and two three-cent pieces; how much money had he?

16. Annie had 60 cents; she spent 30 cents for a doll, and received 10 cents and 2 more from her mother for doing an errand; how many had she then?

17. Thomas had 13 chickens and 11 little turkeys; the cat caught 5 of the chickens and the rats caught 4 of the little turkeys; how many of both were left?

Count by 7's from 1 and 2 to 71 and 72.

18. William had 7 cents and John had twice as many; how many had both the boys?

19. I have 18 pupils in a spelling class; 4 of them misspell some of their words; how many recite perfectly?

---

*The character $ means dollars and $32 is read thirty-two dollars.

20. George had 7 glass agates, 8 china and 6 common marbles ; how many marbles did he have?

21. If 24 apples be equally divided among 8 boys, how many will each have?

22. Walter took 56 cents to the store to buy sugar at 8 cents a pound; how many pounds could he get?

23. Katie's mother gave her 9 cents, her father 7 cents, and her aunt 4 cents; she bought 3 oranges at 4 cents each and spent the rest of the money for candy at 2 cents a stick; how many sticks of candy did she get?

1. 5435	2. 3454	3. 5342
4354	5323	3455
3343	2345	5534
5425	4534	2345
2342	5453	4453
3454	3225	5225
5535	4544	3542
4344	2345	4354
5432	5432	5335
2345	4543	4543
4554	1234	3254
3435	5432	5432
5342	3555	1234
4554	4343	4555
3343	5443	5321

See Teachers' Edition, p. 126.

FIRST STEPS AMONG FIGURES.

4. 3452
   5324
   4543
   5435
   3354
   5445
   4554
   3425
   5432
   2345
   4553
   5434
   4423
   3545
   5434

5. 4532
   3254
   4325
   5543
   3454
   3434
   2345
   5432
   4545
   3454
   5235
   4523
   5445
   4354
   5432

6. 4356
   3645
   5234
   6563
   3656
   4345
   5456
   6563
   3625
   5356
   6545

7. 3564
   6453
   2566
   4635
   5364
   6532
   3656
   4364
   5635
   6556
   3446

8. 6345
   2436
   6563
   3656
   5365
   6534
   4655
   5362
   6656
   3545
   6666

FIRST STEPS AMONG FIGURES.

9.	3456	10.	6354
	6563		5665
	5635		6536
	6366		3456
	3556		6365
	6645		6653
	4536		5546
	5663		3665
	6636		6236
	6355		6463
	5663		3554

11. 684,632−47,757=?   Ans. 636,875.

12. 43,126−2,765=?   Ans. 40,361.

13. 935,643−77,387=?   Ans. 858,256.

14. 754,231−26,154=?   Ans. 728,077.

15. 836,425−68,268=?   Ans. 768,157.

16. 364,135−71,543=?   Ans. 292,592.

17. 463,845−98,536=?   Ans. 365,309.

18. 763,843−46,452=?   Ans. 717,391.

FIRST STEPS AMONG FIGURES.

19. 6453	20. 3456	21. 3645
3566	6566	6556
4635	5635	2663
6326	4364	6356
5663	5662	5565
4566	6556	4636
3656	6636	6353
6365	3665	5664
5636	5326	6656
6656	6553	3533
5625	3466	6665
5564	6536	3456
4563	5462	6363

22. 5663	23. 6543
6656	2345
3546	6234
6635	5623
5366	4562
4653	3456
6636	6566
3565	5335
6366	4666
5436	5656
6563	6535
3666	6666
6525	3553

See Teachers' Edition, p. 126.

When there is a cipher or ciphers in the multiplier between significant* figures, do not use it in multiplying, since nothing times any number is nothing ; but be careful to write the first figure of each product under the figure you multiply by.

1. $57,364 \times 304 = ?$
2. $47,563 \times 504 = ?$
3. $346,752 \times 706 = ?$
4. $630,475 \times 607 = ?$
5. $50,746 \times 467 = ?$
6. $46,375 \times 564 = ?$
7. $3,526 \times 736 = ?$
8. $6,375 \times 657 = ?$
9. $68,574 \times 87 = ?$
10. $47,586 \times 68 = ?$
11. $647,583 \times 85 = ?$
12. $364,758 \times 48 = ?$
13. $105,743 \div 4 = ?$
14. $3,176,207 \div 5 = ?$
15. $1,418,497 \div 4 = ?$
16. $22,477,527 \div 6 = ?$
17. $2,629,369 \div 4 = ?$
18. $23,176,234 \div 5 = ?$

* The significant figures are 1, 2, 3, 4, 5, 6, 7, 8 and 9. See Teachers' Edition, p. 128.

19. 30,249,742 ÷ 4 = ?
20. 2,625,160 ÷ 6 = ?
21. 31,760,850 ÷ 7 = ?
22. 34,459,582 ÷ 6 = ?
23. 34,574,849 ÷ 6 = ?

When the divisor is not contained in the partial dividend, write a cipher in the quotient and the partial dividend will be the remainder to be prefixed to the next figure of the dividend. The teacher will illustrate by the following examples:

24. 10,421 ÷ 4 = ?
25. 44,593,672 ÷ 7 = ?
26. 2,115,885 ÷ 5 = ?
27. 37,548,510 ÷ 6 = ?
28. 25,233,295 ÷ 7 = ?
29. 37,229,621 ÷ 8 = ?
30. 42,519,206 ÷ 6 = ?
31. 40,379,014 ÷ 7 = ?
32. 52,584,387 ÷ 8 = ?
33. 20,326,941 ÷ 8 = ?

1. 874,895 ÷ 201 = ?
2. 728,556 ÷ 201 = ?
3. 12,756,034 ÷ 3,012 = ?
4. 8,604,338 ÷ 2,031 = ?

5. $16,154,855 \div 3,024 = ?$
6. $49,486,701 \div 2,032 = ?$
7. $2,595,960 \div 4,023 = ?$
8. $1,460,703 \div 403 = ?$
9. $10,357,440 \div 3,024 = ?$
10. $14,244,539 \div 3,014 = ?$
11. $13,822,604 \div 4,034 = ?$
12. $25,079,486 - 7,036 = ?$
13. $38,968,336 \div 6,035 = ?$
14. $275,881,300 - 5,046 = ?$
15. $334,654,184 - 70,364 = ?$
16. $349,143,867 \div 6,057 = ?$
17. $226,396,593 - 60,453 = ?$
18. $18,651,776 - 4,076 = ?$
19. $526,026,567 \div 7,046 = ?$
20. $236,326,931 - 5,074 = ?$
21. $39,361,095 \div 6,024 = ?$
22. $540,611,445 - 7,068 = ?$
23. $450,299,132 \div 60,378 = ?$
24. $327,040,029 \div 5,048 = ?$
25. $413,535,496 \div 70,485 = ?$
26. $241,993,12 \div 5,086 = ?$
27. $546,695,551 \div 80,574 = ?$
28. $465,406,942 \div 7,068 = ?$
29. $289,561,188 \div 6,075 = ?$

See Teachers' Edition, p. 136.

FIRST STEPS AMONG FIGURES.

30. 4563
     3656
     6366
     5665
     4656
     6546
     6665
     5656
     3456
     6543
     5655
     6366
     5621

31. 5465
     6536
     6653
     3566
     6632
     5366
     6556
     3465
     6634
     5663
     2556
     6665
     5326

32. 4536
     5665
     4354
     6635
     5666
     6563
     6356
     3665
     5636
     6565
     3663
     6356
     6565

33. 5634
     6565
     3456
     6563
     4656
     5665
     6363
     5636
     3565
     6366
     5636
     6653
     3546

34. 3456
     6536
     5665
     3566
     6343
     4656
     3565
     6666
     5356
     4663
     6536
     5465
     6323

1. A boy had 15 marbles and lost all but six of them; how many did he lose?

2. Mary comes to school 5 days in a week; how many days does she come in 8 weeks?

3. There are 55 sticks of candy in a jar; if 8 little girls each buy a stick, how many sticks will be left in the jar?

4. How many marbles can a boy buy for 27 cents at three cents apiece?

5. Fanny had 8 cents and Julia had 9 cents, how many did both girls have?

When an example involves several operations the pupil should give but one at a time.

6. How much more will 6 oranges cost at 4 cents each, than 7 peaches at 2 cents each?

Solution : If one orange cost 4 cents, 6 oranges will cost 6 times 4 cents or 24 cents. If one peach cost 2 cents, 7 peaches will cost 7 times 2 cents or 14 cents. If the oranges cost 24 cents and the peaches 14 cents, the oranges cost as much more than the peaches as the difference between 24 cents and 14 cents or 10 cents.

7. Henry had 25 cents; he gave 3 cents each to his brother and sister, spent 5 cents for an orange and 2 for candy; how many cents had he left?

See Teachers' Edition, p. 137.

8. Harvey had a twenty-five cent piece, a ten-cent piece, a five-cent piece and a three-cent piece; how much money had he?

9. If I had 4 apples and found as many more, and ate two of them, what part of a dozen had I then? What are they worth at 12 cents a dozen?

10. I have a clock that strikes every quarter hour; how many times will it strike in 9 hours?

11 William spent 12 cents, James spent one third as many and three cents more; how many did James spend?

12. There are 2 little dogs passing; how many eyes, ears and feet have they?

13. Three men each take three bags of wheat to mill, and each bag contained 2 bushels; how many bushels did the men take to the mill?

14. When milk is 6 cents a quart, how many quarts can you get for 42 cents?

15. When milk is 4 cents a quart, how many pints can you get for 20 cents?

16. In a school-room there are 7 rows of seats, and 6 seats in each row; how many seats are there in the room?

17. A lady made 7 squares of patch-work, and her little girl sewed so many that one-half of what both sewed was 10; how many did the little girl sew?

18. How many boxes of wafers at 6 cents a box may be bought for 9 sheets of paper at 2 cents a sheet?

19. How many barrels of apples at $3 a barrel can be given for 6 yards of flannel at $2 a yard?

20. How many four-horse teams can be arranged from 20 horses?

21. Three fields have each 3 trees, under each tree are 3 cows; how many cows in the three fields?

22. A man bought a duck at 9 cents a pound and paid 54 cents for it; how much did the duck weigh?

23. If 6 oranges cost 24 cents, what cost 8 oranges?

Solution: If 6 oranges cost 24 cents, *one* orange will cost one-sixth of 24 cents, or 4 cents, and 8 oranges will cost 8 times 4 cents, or 32 cents.

24. If a boy walks 15 miles in 3 days, at the same rate, how far will he walk in 4 days?

25. If it takes 16 yards of cloth for 2 suits of clothes, how many yards will it take for 6 suits?

26. If a boy goes 8 feet in stepping 4 times, how far will he go in stepping 7 times?

See Teachers' Edition, p. 138.

27. If 3 men can cut 9 acres of grain in one day, how many acres can 6 men cut in a day?

28. If it takes 12 buttons for 3 vests, how many buttons will it take for 8 vests?

29. How many yards of cloth at $3 a yard can be bought for 4 barrels of flour at $6 a barrel?

30. If 3 men can build a wall in 6 days, how long will it take one man?

31. If 5 men can mow a field of grass in 10 days, how long will it take one man?

32. If 4 men cut 8 cords of wood in a day, how many cords will 1 man cut in a day?

33. If 3 men cut a pile of wood in 9 days, how long will it take one man?

34. If 3 mowing machines will cut 27 acres of grass in one day, how many acres will 7 mowing machines cut in one day?

35. If a boy earn 63 cents in 7 days, how much will he earn in 6 days?

36. How many books at 4 shillings each can you buy for 8 dozen eggs at 2 shillings a dozen?

37. A teamster drew 8 loads of stone each day for 7 days; how many loads did he draw?

38. A boy gained 7 cents by selling a knife for 42 cents; what did it cost him?

39. William worked 8 hours at 2 shillings an hour, and Henry worked 3 hours at 3 shillings an hour ; how much did both earn?

40. Four girls have each 2 hens, and each hen has 6 chickens ; how many chickens have the four girls?

Read the following numbers:

1. 750406300.
2. 4576000043
3. 860000307.
4. 15000045001.
5. 7845678437.
6. 37147415006.
7. 47583000000.
8. 370015300.
9. 40000036700.
10. 71000100000.

Write in Arabic the following numbers :
1. Fifteen mil. ten th. three.
2. Two hun. eight bil. one hun. th.
3. Three bil. twenty mil. six.
4. Thirteen bil. nine th. seven hun. forty-five.
5. Ninety-one bil. one mil. one th. one.
6. Four bil. seven hun. fifteen.
7. Two hun. sixty mil.
8. One bil. three hun. sixty mil. two hun. th.

See Teachers' Edition, p. 140.

FIRST STEPS AMONG FIGURES. 89

9. Five mil. ninety.
10. Write in Roman nine hun. thirty-four.
11. Write in Roman seven hun. forty-six.
12. Write in Arabic DCCCXCVII.
13. Write in words 709460371000.
14. Write in Arabic ten th. thirty.

Teach the pupils that the figure at the right expresses units of the first order, the next figure to the left, units of the second order, the next figure, units of the third order, and so on.

15. Write 7 units of the 5th order, 4 of the 3d and 1 of the 1st (in one number.)
16. Write 3 units of the 8th order, 5 of the 7th, 9 of the 3d and 4 of the 2d.

It may aid the pupil in solving the following examples, to put small numbers in the place of the ones given, and see how it would be solved without the slate, then solve in the same manner.

Simple problems for the slate, involving Addition, Subtraction and Multiplication:

1. Mr. Rogers had 746 bushels of wheat and sold 197 bushels of it; how much had he left?
2. George had 295 cents and his father gave him 75 more; how many had he then?

3. Mr. Smith had 96 bushels of oats and Mr. Jones had 9 times as much; how many bushels had Mr. Jones?

4. Lewis has 87 marbles and John has just as many; how many have both boys?

5. Mr. Howard drew 8 loads of oats to market, and there were 79 bushels in each load; how many bushels did he draw to market?

6. From the sum of 79 and 268, take 158.

7. How much will a teacher's salary amount to in 14 years, at $875 a year?

8. James has 47 marbles less than John, and John has 174; how many has James?

9. John lost 15 cents by selling his knife for 90 cents; what did it cost?

10. Miles took 2341 steps in going to school, and Marcus took 560; how many more steps did Miles take than Marcus?

11. Mr. Decker borrowed $150 and paid $65 of it; how much does he still owe?

12. What will 46 bushels of barley cost at 167 cents a bushel?

13. A clerk received a salary last year of $1000. He spent $260 for board and $378 for clothing and other expenses; how much money did he save?

14. There are 30 days in June, and 31 each in July and August; how many days in these three summer months?

See Teachers' Edition, p. 143.

15. There are 168 acres in Mr. Fox's farm, and Mr. Norton's farm contains 89 acres more than Mr. Fox's; how many acres in both farms?

16. A man put $950 in the bank; he drew out $78 at one time, $45 at another, and $159 at another; how much had he left in the bank?

17. John had 39 marbles and Ezra had 13 more than twice as many; how many had Ezra?

18. Mr. Brown bought a farm for $8460 and sold it for $10380; how much did he gain?

19. What cost 369 bushels of oats at 68 cents a bushel?

20. I have 216 bushels of potatoes in 3 bins; there are 59 bushels in one bin and 98 bushels in another; how many bushels in the third bin?

21. A boy having 85 cents, bought a top for 18 cents, a ball for 25 cents, and some oranges for 27 cents; how many cents had he left?

22. If there are 76 bushels of corn in a bin that will hold 950 bushels, how many more bushels of corn may be put into it?

23. A farmer filled at one time 29 bags with oats, and at another 47 bags. If he put two bushels in each bag, how many bushels of oats were put in all the bags?

24. From one million eight hundred thousand take fifteen thousand ninety.

1. 4567	2. 7645	3. 4756
7654	5774	3567
6347	6547	6635
7576	5732	7456
4757	7665	6573
5674	4576	5746
7565	5766	7455
4757	7537	6567
5664	6452	7736
7575	3776	6564
2346	6565	6475

4. 7546	5. 7654
4757	6537
5675	5465
6757	7756
7577	4675
4664	3456
7357	6747
5675	7564
4567	6475
7654	7746
4735	4653

See Teachers' Edition, p. 144.

6. 3456	7. 6754	8. 5746
7653	5673	7457
6575	7566	6575
5747	4375	7667
7356	7647	5734
5676	5774	4757
4747	6757	7676
5635	7577	4567
4567	4652	6735
7476	5767	5657
5745	7766	4576
6574	5473	7465

9. 4576	10. 6457
7654	7564
5773	5735
7566	4657
6757	7576
5674	4657
7365	3456
4537	7573
3456	5746
6765	7757
7576	4567
4457	3456

11. 4,362,516 − 754,359 = ?
12. 736,952 − 78,672 ?
13. 642,534 − 26,356 = ?
14. 352,432 − 86,354 = ?
15. 6,425,314 − 374,321 = ?
16. 463,524 − 71,876 = ?
17. 425,362 − 17,654 = ?
18. 364,253 − 86,174 = ?
19. 463,521 − 186,357 ?
20. 483,654 − 91,987 = ?
21. 635,245 − 12,567 ?
22. 837,524 − 259,286 = ?
23. 43,452,431 − 4,238,865 = ?
24. 756,324 − 85,543 = ?
25. 4,738,536 − 973,659 = ?
26. 86,357 − 7,269 = ?
27. 34,023 − 9,876 = ?
28. 45,300,435 − 2,430,526 = ?
29. 74,200,032 − 5,140,054 = ?
30. 43,250,001 − 5,726,025 = ?
31. 3,400,564 − 210,739 = ?
32. 45,700,325 − 8,730,153 = ?

1. 796,845 × 89 = ?
2. 479,685 × 98 = ?

3. 68,975 × 79 = ?
4. 647,583 × 467 = ?
5. 68,574 × 456 = ?
6. 486,075 × 807 = ?
7. 58,697 × 64 = ?
8. 9,687 × 75 = ?
9. 4,796 × 39 = ?
10. 85,974 × 84 = ?
11. 79,685 × 69 = ?
12. 68,974 × 79 = ?
13. 49,786 × 85 = ?
14. 59,068 × 604 = ?
15. 70,968 × 907 = ?
16. 70,309 × 709 = ?
17. 860,479 × 709 = ?
18. 759,068 × 407 ?
19. 47,096 × 609 = ?
20. 748,609 × 507 = ?
21. 58,709 × 608 = ?
22. 96,047 × 709 = ?
23. 97,806 × 597 = ?
24. 79,689 × 4,759 = ?
25. 896,748 × 6,978 = ?
26. 7,580,065 − 1,251,298 = ?

See Teachers' Edition, p. 145.

27. $3,740,683 - 923,754 = ?$
28. $735,035 - 26,326 = ?$
29. $7,430,246 - 7,503,472 = ?$

1. $3,241,402 \div 7 = ?$
2. $5,078,005 \div 8 = ?$
3. $3,802,457 \div 8 = ?$
4. $49,167,544 \div 9 = ?$
5. $5,076,335 \div 9 = ?$
6. $3,372,081 \div 9 = ?$
7. $3,725,891 \div 8 = ?$
8. $52,301,166 \div 8 = ?$
9. $48,094,605 \div 7 = ?$
10. $43,719,125 \div 9 = ?$
11. $52,629,186 \div 8 = ?$
12. $40,224,713 \div 7 = ?$
13. $67,217,191 \div 9 = ?$
14. $608,226,845 \div 9 = ?$
15. $37,100,695 \div 8 = ?$
16. $460,241,323 \div 7 = ?$
17. $4,781,158,859 \div 6 = ?$
18. $4,885,157,761 \div 7 = ?$
19. $389,158,560 \div 6 = ?$
20. $5,240,869 \div 7 = ?$
21. $3,356,977 \div 7 = ?$

See Teachers' Edition, p. 146.

FIRST STEPS AMONG FIGURES.

22. $6,727,741 - 8 = ?$
23. $5,115,271 \div 8 = ?$
24. $21,367,398 \div 5,024 = ?$
25. $3,722,901 \div 607 = ?$
26. $14,204,241 \div 403 = ?$
27. $158,632,783 - 30,135 = ?$
28. $38,693,395 - 6,024 = ?$
29. $38,464,365 \div 50,396 = ?$
30. $31,970,764 \div 7,048 = ?$
31. $3,891,687,541 \div 60,475 = ?$
32. $2,925,490,533 \div 60,479 = ?$
33. $462,857,740 \div 8,069 = ?$
34. $2,934,401,497 \div 70,586 = ?$
35. $226,663,766 \div 60,379 = ?$
36. $194,513,933 \div 40,297 = ?$
37. $184,110,903 \div 5,048 = ?$
38. $390,761,546 \div 6,037 = ?$
39. $32,688,027,778 \div 70,486 = ?$
40. $674,476,820 \div 9,037 = ?$
41. $27,971,095 \div 6,074 = ?$
42. $355,212,265 \div 5,036 = ?$
43. $23,377,796 \div 5,024 = ?$
44. $452,113,508 \div 7,056 = ?$
45. $299,019,935 \div 406 = ?$

FIRST STEPS AMONG FIGURES.

46. $2,408,592,665 \div 5,064 = ?$
47. $36,979,544 \div 8,026 = ?$
48. $426,696,721 \div 7,014 = ?$
49. $437,791,338 \div 6,038 = ?$
50. $216,526,004 \div 5,027 = ?$
51. $2,464,085,695 \div 604,978 ?$
52. $238,049,090 \div 5,037 = ?$
53. $4,110,929,380 \div 70,586 = ?$
54. $283,899,778 \div 6,034 = ?$
55. $4,980,403,784 \div 70,496 = ?$
56. $2,202,155 \div 463 = ?$
57. $251,864 \div 361 = ?$
58. $434,801 \div 573 = ?$
59. $268,500 \div 463 = ?$
60. $463,799 \div 582 = ?$
61.* $356,116 \div 365 = ?$
62. $357,243 \div 465 = ?$
63. $403,123 \div 586 = ?$
64. $414,617 \div 473 = ?$
65. $4,174,696 \div 485 = ?$

---

* When the left hand figure of the divisor is equal to the left hand figure of the dividend, if the next figure of the divisor be greater than the next figure of the dividend, point off as if the left hand figure of the divisor were greater. The divisor (in one step of the operation) never is contained more than nine times.

See Teachers' Edition, p. 148.

66. 5,499,513÷796=?
67. 5,538,824÷684=?
68. 7,084,249÷896=?
69. 30,734,480÷645=?
70. 43,722,966÷573=?
71. 270,578,240÷4,035=?
72. 268,439,581÷4,657=?

1. William paid 54 cents for 6 doves; what did each dove cost?

2. In an orchard there are 6 rows of trees and 7 trees in each row; how many trees in the orchard?

3. Henry has 8 cents in one pocket and 9 cents in the other; how many cents has he?

4. James has 8 apples and his sister has 6; how many more has James than his sister?

5. What cost 9 knives at 7 shillings each?

6. A boy paid 25 cents for a ball and sold it for 18 cents; how many cents did he lose?

7. If 1 pencil cost 4 cents, what will 8 pencils cost?

8. George bought a knife for 8 shillings, a ball for 5 shillings, and a bat for 2 shillings; what did he pay for all?

9. Marcus spent 8 cents for lemons at 4 cents each; how many lemons did he buy?

See Teachers' Edition, p. 150.

10. If 5 men cut ten cords of wood in a day, how many cords will 7 men cut?

11. If 2 men can dig a certain ditch in 4 days, how long will it take one man to dig it?

12. If 4 men can cradle 12 acres of grain in one day, how many acres will one man cradle in a day?

13. How many rods of wall will one man build in a day, if 3 men build 9 rods in one day?

14. If 3 boys can pick the stones from a meadow in 9 days, how many days will it take one boy to pick them?

15. How many weeks in 35 days?

16. If 4 pounds of sugar cost 36 cents, what cost 8 pounds?

17. When a pineapple costs 18 cents and an orange costs 6 cents, how much more does the pineapple cost than the orange?

18. If a boy can walk 12 miles in 4 hours, how far can he walk in 5 hours?

19. If 4 men can do a piece of work in 8 days, how long will it take one man?

20. Jane bought 5 figs for 3 cents each, and a yard of cloth for 9 cents; how much did she pay for all?

See Teachers' Edition, p. 151

21. Mary sews 4 hours each day, how many hours does she sew in a week?

22. How much will a man's board for a week cost at 4 shillings a day?

23. How much will a man earn in a week, if he gets 9 shillings for a day's work?

24. If 15 cats are on a wall and every third cat jumps off, how many are left?

25. There are 8 quarts in a peck, how many pecks in 32 quarts?

26. How many quarts in 3 pecks?

27. A boy picked 16 quarts of beans and sold them at 25 cents a peck; how much money should he receive?

28. Charles has 7 cents and his brother 3 more than twice as many; how many have both?

29 On Monday morning Mary had 20 sticks of candy; she ate 2 each day, how many had she left the next Monday night?

30. Arthur had 11 peaches, he ate 3 and gave his sister half of the rest; how many did he keep?

31. How many marbles, 2 for 4 cents, can you get for 18 cents?

EXAMPLES FOR THE SLATE.

If the pupil will use small numbers instead of the large ones in the following examples, and think carefully how he would solve them if they were not for the slate, and then do the same with the numbers given, using the slate as a help, he will be greatly assisted.

1. A man had $3,210, he spent $978 for wheat and $749 for corn; how much money had he left?

2. What is the product of 9,687 and 75?

3. From the sum of 3796 and 4279, take their difference?

4. If a farmer have 256 gallons of cider, how many barrels holding 36 gallons can he fill?

5. From a cistern holding 743 gallons, 98 gallons were pumped out and afterwards 39 gallons poured in; how many gallons were then in the cistern?

6. What cost 37 carriages at $185 each?

7. If a ship sail 7289 miles in 37 days, how many miles does she sail per day?

8. A miller paid $169 for 78 bushels of wheat, $97 for oats and $395 for corn; what did he pay for all of the grain?

9 From the sum of 397 and 6798, take 69.

See Teachers' Editon, p. 152.

10. The difference between two numbers is 347, and the less number is 79, what is the greater number?

11. A man died leaving $5600, of which he gave his wife $2,800, his son $900, one daughter $850 and the rest to another daughter; how much did the second daughter receive?

12. A man bought 75 sheep at one time, and 169 at another; he sold 86 of them to one man and 49 to another; how many had he left?

13. Mr. Wilson bought one house for $4150, and afterward another for $3750; he sold both of them for $7000; did he gain or lose, and how much?

14. There is an orchard consisting of 24 rows of trees, and 36 trees in each row; how many apples in the orchard, allowing an average of 2079 on a tree?

15. A man owing $7165, gives in payment 39 cows valued at $48 each and $750 in money; how much does he still owe?

16. Add 16 thousand 20, fifty millio. 1 thousand nine, 79 thousand 847, and 9 m. lion 79 thousand 8.

17. How many tons of hay at $18 a ton must be given for 16 horses at $153 each?

18. $639 + 91,758 + 9,347 + 81,731 + 9,342 + 35,446 + 8,237 + 12,849 + 87,677 = ?$

19. A grocer spent $881 for molasses and sugar; he paid $368 of the money for molasses, and the rest for 27 barrels of sugar; how much did the sugar cost a barrel?

20. How many yards of cloth in 68 bales, each bale having 97 pieces, and each piece containing 29 yards?

21 Paid $6 each for 75 sheep, and sold the flock for $400; did I gain or lose, and how much?

22. How many horses at $165 each can be bought for $2360?

23. How much is gained by buying 48 cows at $37 each, and selling them at $45 each?

24. Mr. Dixon has 225 acres of land worth $97 an acre, and Mr. Taft has 196 acres worth $79 an acre; how many acres have the two together, and what is the value of the whole?

25. A man sold a farm of 96 acres at $9 an a..., and with the money received for it ...nt a farm of 135 acres; what did he ..y an acre for the latter farm?

26. A teacher had his life insured for $2500. At the time of his death he owned a house and lot worth $1850 and furniture worth $475. He owed debts to the amount

of $369; how much did he leave his family?

27. There are 5280 feet in a mile; how many feet in 709 miles?

28. A man starts from New York on Tuesday morning and travels at the rate of 57 miles a day; another starts from the same place Wednesday morning and follows on at the rate of 69 miles a day; how far apart are they Thursday night?

29. James sold a grocer 96 eggs at 15 cents a dozen, and received 120 cents; how much does the grocer still owe him?

30. If there were 365 days in each year, how many years would there be in 31390 days?

31. Add seventy million nine hundred thousand, two hundred six thousand eight, sixty thousand sixty, seven thousand nine hundred, ten million ten thousand ten, and seven hundred fifty-nine million two hundred thirty thousand.

1. 7657	2. 3456	3. 7567
4775	5767	4756
7777	7475	3675
3456	4567	6777
6735	7756	7546
5677	5647	5735
7756	6775	6767
3457	7564	7476
6574	4677	3456
5767	3456	3456
4575	5767	7763
7757	7534	4567
5676	6467	2475
2345		

4. 4756	5. 7654
7577	4775
6645	5467
7734	7556
5675	6775
6756	3457
7467	5747
4575	7674
5647	6577
7777	7754
6452	7777
3567	7777
6776	4564
5643	5675

See Teachers' Edition, p. 153.

6.	4568	7.	6758	8.	8765
	3785		8547		4658
	8678		7868		7777
	5786		8778		3456
	8678		8888		6778
	8888		8888		7567
	8888		3576		8878
	5678		6758		7767
	8765		8687		8585
	4876		5678		7777
	5487		6786		3456

9.	8888	10.	3748
	8888		8675
	7654		7887
	4765		4567
	7777		8888
	7777		8888
	5678		4567
	8765		7777
	4567		7777
	8778		4565
	7658		8486

11. 6758
    7584
    4676
    8778
    3456
    8888
    8888
    6547
    3754
    6678
    7563
    5885

12. 7658
    4576
    6785
    5467
    8878
    4657
    8765
    6578
    7857
    5686
    8578
    6785

13. 7865
    4578
    8657
    6785
    8888
    8888
    3456
    7777
    5678
    8753
    4576
    8687

14. 4786
    8657
    6578
    7865
    4576
    8888
    3456
    7777
    8765
    3578
    8657
    4768

15. 7684
    4578
    3456
    8888
    8888
    7654
    3567
    7777
    5648
    7385
    4637
    8386

See Teachers' Edition, p. 154.

16.	17.	18.
3687	4837	4587
8546	7584	8635
7685	8758	6754
4868	5875	8888
5784	8888	8888
7777	8888	6754
6548	6753	7578
3675	4584	4785
8888	7777	7777
8888	5678	5768
5678	8765	8654
8563	3857	4585
5785	6586	7848
8678	8465	5686

19.	20.
8476	7586
5768	4767
4567	8658
7777	4875
7777	8888
8654	8888
3568	7654
8888	3867
8888	7777
4567	4685
8765	8568
4478	7777
7586	4825
8857	8674

1. 4,570,365 − 323,456 = ?
2. 9,374,056 − 636,587 = ?
3. 685,700,365 − 296,314,537 = ?
4. 76,400,235 − 3,234,567 = ?
5. 38,500,684 − 8,769,876 = ?
6. 7,460,683 − 379,876 = ?
7. 375,600,735 − 83,735,829 = ?
8. 83,640,574 − 5,712,653 = ?
9. 794,600,435 − 63,732,367 = ?
10. 74,300,375 − 53,620,547 = ?
11. 8,750,043 − 970,236 = ?
12. 6,713,021 − 6,873,213 = ?
13. 48,300,563 − 9,000,687 = ?
14. 487,500,564 − 65,730,637 = ?
15. 756,000,375 − 85,203,456 = ?
16. 6,847,000,346 − 367,020,654 = ?
17. 79,068 × 58 = ?
18. 80,479 × 74 = ?
19. 4,185 × 368 = ?
20. 968,579 × 798 = ?
21. 79,689 × 4,759 = ?
22. 79,867 × 6,897 = ?
23. 85,765 × 81,072 = ?
24. 49,678 × 9,876 = ?
25. 497,896 × 8,659 = ?

See Teachers' Edition, p. 154.

Since moving a figure one place to the left increases its value ten fold, and moving it two places, ten times ten fold or one hundred fold,—to multiply any number by 10, 100, 1000, &c., annex as many ciphers to the multiplicand as there are in the multiplier.

$379 \times 100 = 37,900$ **Ans.**
26. $7,865 \times 10,000 = ?$
27. $573 \times 10 = ?$
28. $68 \times 100 = ?$
29. $6,320 \times 1000 = ?$
30. $875 \times 100,000 = ?$

When there are ciphers at the right of either the multiplier or multiplicand, or of both, place the multiplier under the multiplicand so that the significant figures farthest to the right shall come under each other. After multiplying by the significant figures and adding, write as many ciphers at the right of the product as there are at the right of the multiplier and multiplicand together.

(These directions are given very minutely but are not to be committed to memory.)

$$34200$$
$$34000$$

$$1368$$
$$1026$$

$$1{,}162{,}800{,}000$$

31. $3{,}750 \times 46{,}000 = ?$
32. $46{,}300 \times 350 = ?$
33. $635{,}000 \times 700 = ?$
34. $27{,}500 \times 680{,}000 = ?$
35. $586{,}000 \times 7{,}400 = ?$
36. $490 \times 36{,}700 = ?$
37. $6{,}840 \times 7{,}500 = ?$
38. $8{,}609 \times 800 = ?$
39. $67{,}900 \times 870 = ?$
40. $8{,}690 \times 4{,}700 = ?$
41. $480{,}600 \times 7{,}090 = ?$
42. $70{,}580 \times 6{,}408{,}000 = ?$
43. $706{,}900 \times 5{,}078{,}000 = ?$
44. $68{,}090 \times 70{,}900 = ?$
45. $640{,}980 \times 10{,}000 = ?$

1. $43{,}188{,}278 \div 9 = ?$
2. $791{,}071{,}117 \div 9 = ?$

3. $6,156,712,635 \div 8 = ?$
4. $67,815,232 \div 7 = ?$
5. $6,928,288,028 \div 9 = ?$
6. $52,717,437 \div 8 = ?$
7. $8,808,273,807 \div 9 = ?$
8. $461,392,186 \div 6 = ?$
9. $49,347,765 \div 7 = ?$
10. $44,160,343 \div 9 = ?$
11. $557,793,576 \div 7 = ?$
12. $7,664,063,843 \div 8 = ?$
13. $8,044,185,607 \div 9 = ?$
14. $461,093,406 \div 7 = ?$
15. $5,225,741 \div 6 = ?$
16. $529,762,735 \div 7 = ?$
17. $7,230,245 \div 8 = ?$
18. $86,274,817 \div 9 = ?$
19. $716,863,843 \div 8 = ?$
20. $41,088,317 \div 7 = ?$
21. $30,884,751 \div 7,058 = ?$
22. $34,600,073 \div 6,032 = ?$
23. $1,890,186 \div 5,178 = ?$
24. $1,203,161,896 \div 8,169 = ?$
25. $5,279,490 \div 814 = ?$
26. $33,620,328 \div 725 = ?$

See Teachers' Edition, p. 155.

27. 92,197,364÷6,257=?
28. 7,638,482÷439=?
29. 511,764,908÷7,461=?
30. 946,526,656÷6,397=?
31. 65,790,555÷6,847=?
32. 6,363,666÷469=?
33. 18,827,247÷378=?
34. 526,493,286÷3,794=?

Since moving a figure one place to the right diminishes its value ten fold, and two places, ten times ten fold or one hundred fold,—to divide any number by 10, 100, 1000, &c., cut off by a vertical line as many figures on the right of the dividend as there are ciphers at the right of the divisor.

The number at the left of the vertical line will be the quotient, and the number at the right of it the remainder.

Illustration : 78634÷100=?

Solution : 786 | 34 the quotient is 786 and 34 is the remainder.

1. 793,468÷10,000=?
2. 37,680÷100=?
3. 2,347,600÷100,000=?
4. 76,219,648÷100=?
5. 372,938,641÷10,000,000=?

To divide by any number with ciphers at the right.

$$78{,}673 \div 700 = ?$$

Divide both dividend and divisor by 100, and cutting off the 2 figures at the right, and the example becomes—

7|00)786|73

112—273 rem., or

$$\begin{array}{r}273\\112\underline{}\\ \cdot\ 700\end{array}$$

Divide the number at the left of the vertical line in the dividend, by the number at the left of the vertical line in the divisor, and to the remainder annex the figures of the dividend cut off.

45|000)612|370(13 quo.

$$\begin{array}{r}45\\ \underline{}\\ 162\\ 135\\ \underline{}\\ 27370\ \text{rem.}\end{array}$$

Hence the rule: To divide by any number with ciphers at the right, cut off the ciphers at the right of the divisor by a vertical line, and also as many figures at the right of the dividend. Divide the remaining number in the dividend by the remaining number in the divisor, and to the remainder annex the figures cut off from the right of the dividend for the true remainder.

1. $18,228,211 \div 37,500 = ?$
2. $5,142,762,131 - 750,000 = ?$
3. $546,927,300 \div 687,000 = ?$
4. $75,514,192 \div 796,800 = ?$
5. $8,734,758 \div 10,000 = ?$
6. $350,870,000 \div 3,580,000 = ?$
7. $3,278,300 \div 7,000 = ?$
8. $87,200 \times 23,000 = ?$
9. $7,162,323 - 900 = ?$
10. $394,690,750 - 5,800 = ?$
11. $29,850,010 \div 3,750 = ?$

12. From six billion six thousand six, take eighty million eight.

13. $27,752,320,172 \div 570,000 = ?$
14. $2,910,144,700 \div 36,800 = ?$

See Teachers' Edition, p. 155.

15. 178,576,495 ÷ 100,000 = ?
16. 365,820,038 ÷ 6,000 = ?
17. 475,308,056 ÷ 48,600 = ?

18. Subtract forty-five million, one thousand ten, from forty-two million seven hundred thousand.

19. 441,937,000 ÷ 597,000 = ?
20. 30,500,857,231 ÷ 3,780,000 = ?
21. 3,657,200 ÷ 600 = ?
22. 47,096 × 8,600 = ?
23. 70,286,631 ÷ 900 = ?
24 7960 × 100 = ?

When the multiplier is less than 13 the pupil should be taught and *required* to multiply but once through, multiplying by 11 or 12 as he has already been taught to multiply by 4, 5 or 6.

1. 78,967 × 12 = ?
2. 69,789 × 11 = ?
3. 754,836 × 12 = ?
4. 845,768 × 12 = ?
5. 68,094,796 × 11 = ?
6. 586,974 × 12 = ?

See Teachers' Edition, p. 157.

7. 6,579,687 × 12 = ?
8. 69,487,968 × 12 = ?
9. 9,468,579 × 11 = ?
10. 479,658 × 12 = ?
11. 97,589,647 × 12 = ?
12. 6,975,897 × 12 = ?
13. 69,786,995 × 12 = ?
14. 94,679,689 × 12 = ?
15. 979,896 × 12 = ?
16. 685,796 × 11 = ?
17. 8,798,979 × 12 = ?
18. 49,897,697 × 12 = ?
19. 97,987,986 × 12 = ?
20. 7,989,985 × 12 = ?
21. 1,185,491,377 ÷ 12 = ?
22. 546,569,461 ÷ 11 = ?
23. 820,499,872 ÷ 12 = ?
24. 1,044,922,315 ÷ 12 = ?
25. 9,438,575,969 ÷ 12 = ?
26. 9,568,867,509 ÷ 11 = ?
27. 455,976,730 ÷ 12 = ?
28. 956,219,625 ÷ 12 = ?
29. 898,436,885 ÷ 12 = ?
30. 7,434,306,992 ÷ 11 = ?

FOR ORAL RECITATION.

1. A boy having 25 cents, bought marbles at 4 cents each, keeping 5 cents of the money ; how many marbles did he buy?

2. Jane lost 10 cents on her way to the post-office, and spent the rest of her money for 10 3-cent stamps ; how much money had she when she started?

3. * Mark earned 8 cents, lost 5 cents, and then found 10 cents, when he had 25 cents ; how much money had he at first?

4. What cost 12 pounds of sugar if 7 pounds cost 63 cents?

5. A man spent $5, then earned $7, and after giving away $6 found he had $15; how many dollars had he at first?

6. If 6 apples cost 2 cents, what cost 18 apples?

Solution : If 6 apples cost 2 cents, 18 apples, which are 3 times 6 apples, will cost 3 times 2 cents or 6 cents.

7. If 4 marbles cost 3 cents, what cost 24 marbles?

---

* If he had 25 cents *after* finding 10 cents, before he found it he had the difference between 25 cents and 10 cents, or 15 cents. If he had 15 cents *after* losing 5 cents, *before* he lost it he had the sum of 15 cents and 5 cents or 20 cents. If he had 20 cents *after* earning 8 cents, before he earned it he had the difference between 20 cents and 8 cents or 12 cents.

See Teachers' Edition, p. 161.

8. What cost 20 figs, if 5 figs cost 2 cents?

9. If 3 oranges cost 12 cents, what cost 7 oranges?

10. What cost 12 lemons, if 5 cost 25 cents?

11. How many apples can be bought for 15 cents, at the rate of 5 for 3 cents?

12. If 3 oranges cost 10 cents, how many may be bought for 40 cents?

13. 30 cents will buy how many apples, at 9 for 6 cents?

14. How many figs may be bought for 24 cents, at the rate of 3 figs for 2 cents?

15. If 4 marbles cost 5 cents, what cost 20 marbles?

16. If 5 lemons cost 20 cents, what cost 9 lemons?

17. What cost 30 pears if 3 pears cost 5 cents?

18. If 4 peaches cost 3 cents, what will 24 peaches cost?

19. At the rate of 2 oranges for 9 cents, how many may be bought for 18 cents?

20. If 3 men cut 6 cords of wood in a day, how many cords will 7 men cut in a day?

21. If 3 men dig a ditch in 12 days, how long will it take one man?

See Teachers' Edition, p. 162.

22. If 4 men harvest a field of wheat in 8 days, how many days will it take 1 man to harvest it?

23. If 12 men can dig a field of potatoes in 13 days, how many men will do it in 1 day?

24. How many men can load a car in 1 hour, if 2 men can load it in 4 hours?

25. If 4 men can do a piece of work in 12 days, how long will it take 3 men to do it?

26. How many days will it take 6 men to earn $32, if it takes 4 men 6 days to earn it?

27. If 4 men can do a piece of work in 9 days, how many men can do it in 6 days?

28. If 6 men can do a piece of work in 4 days, how many men will it take to do the work in 3 days?

29. If 6 men can do a piece of work in 12 days, how long will it take 4 men to do it?

30. If 4 men can build 12 rods of wall in a day, how many rods can 6 men build in a day?

31. How many men will build a wall in 12 days, if 6 men build it in 8 days?

32. If 3 men cut 7 cords of wood in a day, how many cords will 12 men cut in a day?

33. A girl took 7 pins from a paper and then put on 9; her brother afterwards took off 6, and left in it 24; how many on the paper at first?

34 How many men will do a work in 6 days that 9 men do in 4 days?

35. If 8 men do a work in 6 days, how many men will do it in 12 days?

36. If 4 men do a work in 12 days, how long will it take 6 men?

37. If 3 pencils are worth 11 cents, how many pencils can be bought for 33 cents?

38. A girl having a paper of candy, ate 7 pieces; then her brother gave her 5 pieces, after which she gave her mother 9 pieces. She had left 27 pieces; how many pieces had she at first?

39. What number divided by 2 will give 6?

40. If 4 cords of wood cost $20, how many cords can be bought for $35?

41. If 6 vests are worth $24, what are 9 vests worth?

42. If 5 cords of wood cost $24, what will 15 cords cost?

43. What number divided by 3 will get 12?

44. At 10 cents a pint, what will a gallon of molasses cost?

45. How many bushels of potatoes at 4

shillings a bushel may be bought for 3 bushels of wheat at 12 shillings a bushel?

46. A boy gave 8 marbles worth 2 cents apiece, for 7 pencils worth 3 cents each; how much did he gain?

47. How many eight-gallon cans will be required to hold 56 gallons of milk?

48. If 15 bushels of wheat will make 3 barrels of flour, how many bushels will make 8 barrels?

49. How many yards of cloth at $6 a yard will pay for 9 tons of coal at $8 a ton?

50. When flour is $7 a barrel, how many barrels can be bought for $8, and 9 bushels of wheat at $3 a bushel?

51. 96 eggs are how many dozen?

52. If 8 horses eat 48 bushels of oats in 2 weeks, how many bushels will 5 horses eat in the same time?

53. Our school has a recess in the forenoon and also in the afternoon. If there are one hour of school before each recess and two hours after each recess, how many hours of school in a week?

54. If two apples cost one-half of 10 cents, how many can be bought for 15 cents?

55. How many three-cent stamps can be bought for 27 cents?

56. A boy caught some fishes; he threw

away 7, then caught 3 and bought 2, when he had 14; how many did he catch at first?

57. Kate lives two miles from school, and does not go home at noon; how far must she walk in a week if she loses no time at school?

58. Frank has 6 five-cent pieces, 4 three-cent pieces and five two-cent pieces; how many cents has he?

59. John has 7 cents, his brother 8, and their sister has 4 more than both of them; how many have they all?

60. Which costs the more, 3 lemons at 4 cents each or 6 pears at 2 cents each?

61. A boy went to the grocery with 25 cents, and bought 2 pounds of sugar at 9 cents a pound; how much change should he bring back if he has 2 cents for doing the errand?

62. If 2 barrels of flour will last 3 men 6 months, how long will it last 9 men?

63. Bought some peaches for 24 cents, at the rate of 5 for 2 cents, and divided them equally among 6 boys; how many did each boy receive?

64. If 7 bushels of clover seed are worth $42, how many bushels of wheat at $2 a bushel will 3 bushels of clover seed buy?

65. If 2 men start from the same place and travel in the same direction, one 6 miles an hour, and the other 3 miles an hour, how far apart will they be in 9 hours?

66. In how many hours will a man who drives 8 miles an hour overtake a footman who is 60 miles ahead, and walks at the rate of 3 miles an hour?

67. A man bought a span of horses for $100, paid $60 for their keeping, and sold them for $200; what did he gain on each horse?

68. How many turkeys can I buy for $43, at the rate of 3 for $5, and have $8 left?

EXAMPLES FOR THE SLATE.

1. In a certain church 28 pews rent at $35 each, 19 at $25 each and 37 at $15 each; for how much do they all rent?

2. A railroad 18 miles long cost $452,682 for labor, and $177,228 for other expenses; what was the cost per mile?

3. One half of the inhabitants of Constantinople are Turks, 150,000 Greeks, 30,000 Armenians, and 65,000 Jews; how many in all?

See Teachers' Edition, p. 165.

4. If a man earns $960 a year, and spends yearly $688, in how many years will he lay up $4,624.

5. A man bought a farm for $17,600; he sold half of it for $9,322, at the rate of $79 an acre; how many acres did he buy? How much did he pay an acre?

6. From the sum of 7574 and 10746, take their difference.

7. A lady having $125, paid $37 for a set of furs, and $2 a yard for 23 yards of silk; how much money had she left?

8 133416 emigrants arrived in New York in 1867, which was 9,731 more than arrived in 1866; how many arrived in 1866?

9 James and George started together, and traveled in the same direction. James walked 2 miles an hour and George 4 miles an hour; how far apart were they at the end of 19 hours?

10. In six boxes of crayons there are 864 pieces; if 864 pieces cost 360 cents, what will one box cost?

11. There are two numbers, the greater of which is 37 × 96, and their difference is 18 × 27; what are the numbers?

12. A earns $45 a month, and B earns 13 times as much lacking $490; how much does B earn in 8 months?

13. If a house is worth $1,800 and the farm on which it stands five times as much lacking $36, and the stock one-third as much as the house and farm, what is the value of the whole?

14. A man sold his farm of 245 acres at $69 an acre and bought some land at $97 an acre; how many acres could he buy?

15. Mr. Smith was 968 miles from home; he traveled toward home 137 miles one day; 119 the next day, and 98 the third day; how far was he from home then?

16. From thirty billion ten thousand, take seven billion two hundred nine thousand seventy-five.

17. A man bought 325 bushels of barley for $500; 450 bushels of oats for $250; 625 bushels of corn for $150 more than he paid for the oats; 300 bushels of beans at $2 a bushel, and some wheat for $100 more than he paid for the corn; how much did he pay for all?

18. How many solid feet of earth can be removed in 36 days by two carts each carrying 9 loads a day, and 34 solid feet at a load?

19. A man having $9,840, gave each of his two sons $2,750 and the remainder to his daughter; how much did he give his daughter?

20. $26,250 is 3 times what A gave for his farm, and he gave $370 more for it than it was worth; how much was the farm worth?

21. I sold a horse for $375 which cost me $295; how much did I gain?

22. I sold a cow for $65 and by so doing lost $15; what did she cost?

23. A man began business with $3,850, and in 7 years he was worth $10,465; how much did he make each year?

24. How many days would 36 horses live on an amount of food that would keep 24 horses 288 days?

25. A merchant received $248 on Monday and $396 on Tuesday; what was the average receipts per day?

26. Two men start from the same place and travel in opposite directions, one at the rate of 54 miles a day, and the other at the rate of 45 miles a day; how far apart will they be at the end of 6 days? How far apart if they travel in the same direction?

27. A man bought 478 bushels of corn; all but 136 bushels were sunk in a boat; how much was saved?

28. A merchant bought 46 yards of cloth for $93, and sold it at $3 a yard; how much did he gain?

See Teachers' Edition, p. 166.

29. Divide the product of 79 and 237 by their difference.

30. At $135 each, how many horses can be bought for $9,368?

31. How many times can 317 be subtracted from 13,314?

32. There are 3 bins containing 856 bushels of wheat; 1 contains 376 bushels, another contains 297 bushels; how many in the third bin?

33. A farmer sold 13 tons of hay at $16 a ton, and 24 cords of wood at $5 a cord; he divided the money received among four creditors; how much money did each receive?

34. A has 18 barrels of flour of 196 pounds each; if a family of 9 persons use 49 pounds of flour a week, how long will the flour last them?

35. If Mr. Long's sheep were put into 6 fields, 96 in a field, there would be 5 sheep remaining; how many sheep has he?

36. A grocer bought 2 cheeses, one weighing 68 pounds and the other 75 pounds, at 14 cents a pound; how many cents would he gain by selling them at 17 cents a pound?

37. A man killed four hogs, one weighing 368 pounds, one 412, one 379 and one 433; what was their average weight?

38. There were 84 sheep in four pastures; there were 30 in the first and 24 in the second; if there were an equal number in each of the others, how many in each?

39. If a man paid $500 for four horses, $200 for 5 cows and $175 for 40 sheep, how many animals did he buy?

40. If a man earns $685 a year, and spends $496 a year, in how many years will he save $1,134?

41. How many pounds of coffee at 27 cents a pound will pay for three hogsheads of sugar, each containing 1080 pounds, at 12 cents a pound?

42. What is the sum of the difference and sum of 1768 and 987?

43. A man deposited in bank at different times $397, $450 and $568; he drew out at one time $275 and at another $368; how much remained in the bank?

44. A man sold 26 cows at $35 each; how many horses at $145 each can he buy with the money received?

45. A dealer shipped 500 bushels of beans in 250 bags, 600 bushels of wheat in 280 bags; he used 136 less bags in which to ship 300 bushels of corn than he did for the wheat; he put 1200 bushels of oats in bags holding 2 bushels each; how many bags did he use for all the grain?

46. The income of a man who "struck oil" is $75 a day; how many teachers would this employ at $850 a year?

47. A farmer having $1397, bought 9 tons of hay at $16 a ton, a horse for $185, 155 sheep at $4 each, and spent the rest of his money for cows at $32 each; how many cows did he buy?

48. A fisherman caught 2 dozen fishes; he sold one-half of them at 25 cents each; the other half for 26 cents each, except one which, weighing 33 pounds, he retailed at 11 cents per pound; how much did he get for his fishes?

49. A boy paid 100 cents for 5 quires of paper (24 sheets each) and sold it at the rate of 2 sheets for 3 cents; did he gain or lose, and how much?

50. How many half dimes in 350 cents?

51 A miller ground 34 bushels of wheat, 18 of corn, and 22 of oats; how many bags holding 2 bushels each, held the grain? What did the grinding cost at 7 cents a bushel?

52. 24 boys attended the same school, but in three different rooms; 5 were in one room, and 8 in another, and if the number of boys in the third room be multiplied by 12, the product will equal the number of blackbirds they saw on their way to school; how many did they see?

53 Sound travels at the rate of 1090 feet in a second; at this rate how long would it take the report of a cannon to reach the moon, which is 240,000 miles away (1 mile is 5280 feet)?

54. An estate of $14350 was divided between a widow and two children; the widow's share was $5450, the son's $1280 less than the widow's, and the daughter had the rest; how much did the daughter have?

55. The product of two numbers is 36288, and one of them is 756; what is the other?

56 A man bought 145 acres of land for $9,850, and 95 more acres at $45 an acre; he sold the whole at $56 an acre; did he gain or lose, and how much?

57. A farmer bought 47 acres of land for $4,416, and 34 acres at $75 an acre; what was the average price per acre?

58. The sum of two numbers is 7568, and one of them is 784; what is the other?

59 How many military companies of 98 men each, can be formed from 7,463 men?

60. How many yards of cloth at 24 cents a yard, will pay for 26 dozen eggs at 14 cents a dozen, and a jar of butter worth 284 cents?

61. George and Lewis start from the same place at the same time, and travel in the same direction, George at the rate of 714 rods an hour, and Lewis at the rate of 579 rods an hour; how far apart are they at the end of 9 hours? How far apart in 7 hours, if they had traveled in opposite directions?

62. If I receive $40 a month and spend $32 a month, in how many years will I save $1,152?

63. Subtract the difference between 79 and 2300 from their sum.

64. What is the sum of ten thousand ninety, seven thousand nine hundred, eight million nine hundred eighteen, five hundred thousand, seventy thousand seventy-five, and eight hundred.

65. The dividend is 736592, the divisor is 6978; what is the quotient and remainder?

66. The remainder is 658 and the subtrahend 1734; what is the minuend?

Pupils make and solve the following examples:

67. Given a multiplicand of 4 figures, a multiplier of 3 figures, required the product?

68. Given the minuend and remainder, find the subtrahend.

69. Given the subtrahend and the remainder, find the minuend.

70. Given the sum of three numbers and two of them, find the third.

71. Given the difference between two numbers and the less number, find the greater.

72. Given the divisor, quotient and remainder, find the dividend.

73. Given the product of two numbers and one of them, find the other.

74. Given the difference between two numbers and the greater number, find the less number.

75. Given whole price, number of articles, find the price of a different number of articles.

76. Given the cost and selling price, find the gain.

77. Given the selling price and the loss, find the cost.

78. Given the cost and the gain, find selling price.

79. Given the selling price and gain, find cost.

80. Given the cost and the loss, find the selling price.

FIRST STEPS AMONG FIGURES.

1. 6745	2. 5847	3. 9337
5678	9576	4598
9867	4684	3765
6543	5967	9458
7698	8439	7695
4759	4785	4739
5978	9478	5345
9647	4567	4869
8458	8975	3765
4796	4687	4798

4. 8451	5. 5747
6759	9835
7846	4696
9567	8739
6976	5684
7569	7938
8427	6456
9568	8845
3753	7587
5679	8946

6. 5768	7. 4768	8. 5869
9576	9535	9768
4869	9849	4352
8753	7697	4675
4968	4857	9538
7495	9564	4678
6478	5739	5765
4956	5647	4976
9875	9788	9738
5637	5347	4657
8592	8234	9876

9. 7648	10. 7465
3752	5876
9746	4795
3859	8649
8432	3578
9594	9756
3745	6549
9458	8732
4637	5685
8358	7839
3543	5693

11. 4538
     9764
     3797
     8436
     9768
     4593
     8945
     7637
     4895
     7537
     4885
     5938
     9457

12. 9638
     5796
     4579
     8947
     6896
     9758
     4563
     9755
     6879
     5768
     8597
     4856
     9589

13. 9768
     7596
     4879
     9687.
     4739
     8645
     3787
     9568
     4837
     8795
     4568
     9756
     6897

14. 6795
     4569
     8769
     9345
     8876
     6789
     3954
     8456
     3789
     8375
     9999
     3478
     6457

15. 5896
     9748
     7635
     5864
     9787
     4538
     7687
     4859
     5321
     4978
     9654
     7987
     9868

16. 6587	17. 9768	18. 4879
4759	4593	9763
5896	7846	4598
9537	5937	8756
4678	9876	4975
4957	2345	6874
8436	3527	7589
6758	8498	4837
4957	6549	5692
8796	7856	7859
4587	7387	4537
7948	9765	6895
6595	3849	4576
4769	9674	3844
9487	3758	7989
5896	6847	4596

FIRST STEPS AMONG FIGURES. 139

19.	4532	20.	9476
	7856		3869
	4978		5458
	9456		7567
	3279		4835
	4856		8769
	7995		7654
	8447		8579
	9568		6432
	4789		4976
	6435		9845
	7896		5637
	4967		7948
	8538		8654
	7689		6739
	9876		4798

www.ingramcontent.com/pod-product-compliance
Lightning Source LLC
Chambersburg PA
CBHW021155230426
43667CB00006B/409